CORROSION PROCESSES

CORROSION PROCESSES

Edited by

R. N. PARKINS

B.Sc., Ph.D., D.Sc., F.I.M.

*Professor, Department of Metallurgy and Engineering Materials,
The University of Newcastle upon Tyne, UK*

APPLIED SCIENCE PUBLISHERS
LONDON and NEW YORK

APPLIED SCIENCE PUBLISHERS LTD
Ripple Road, Barking, Essex, England

Sole Distributor in the USA and Canada
ELSEVIER SCIENCE PUBLISHING CO., INC.
52 Vanderbilt Avenue, New York, NY 10017, USA

British Library Cataloguing in Publication Data

Corrosion processes.
1. Corrosion and anti-corrosives.
I. Parkins, R. N.
620.1'623   TA462

ISBN 0–85334–147–8

WITH 33 TABLES AND 124 ILLUSTRATIONS

© APPLIED SCIENCE PUBLISHERS LTD 1982

Photoset in Malta by Interprint Limited
Printed in Great Britain by Galliard (Printers) Ltd, Great Yarmouth

PREFACE

The alleged costs of corrosion in a number of industrialized countries where estimates have been made amount to something in the region of 4% of the gross national product. Although such calculations may be open to objection on matters of detail, the magnitude is likely to remain high whatever route is followed to arrive at the final cost. This, coupled with the visible manifestations of corrosion in everyday life spanning domestic and industrial usage of metals, provides sufficient incentive for continuing research and development into understanding the causes of corrosion and methods for its control.

The fields of research that cover corrosion and its prevention are consequently in a continuous state of flux, with the literature published in learned journals increasing exponentially with time and the problems associated with keeping abreast of developments therefore difficult, as they are in many fields of science and technology. One of the solutions to this dilemma is books that review the situation with regard to particular topics: hence the present volume.

Of course, the contents of such volumes reflect the subjective assessments of the editor and the authors, even though the latter are chosen because of their expertise in particular areas. Different facets of a subject develop at different rates in a given time period. For example, general corrosion of metals, which economically is probably still by far the most important area in this field, appears to have attracted fewer published papers in the last decade than the localized forms of corrosion. The emphasis in the present volume is more on the forms of corrosion and an understanding of their causes, rather than on preventive measures, although the latter are implied or dealt with specifically, even if briefly, in the various chapters.

Chapter 1 reviews recent applications of electrochemical techniques to atmospheric corrosion research and to monitoring of atmospheric corrosion phenomena. Some difficulties in interpreting time-of-wetness measurements from electrochemical sensors are pointed out and suggestions to use a time-of-wetness and a corrosion time or to determine the times for which adsorbed or phase layers are present are discussed. Finally, the chapter discusses attempts to make predictions concerning long-term atmospheric corrosion behaviour based on electrochemical measurements.

In Chapter 2, international studies of atmospheric corrosion of non-ferrous metals published over about 25 years have been reviewed. Types of atmospheres and the factors affecting metallic corrosion are detailed; outdoor and accelerated corrosion testing procedures are briefly discussed.

There is treatment of the corrosion performance of aluminium, magnesium, zinc, cadmium, copper, tin, lead and nickel and their alloys and how the various factors affect their atmospheric corrosion performance.

Chapter 3 deals with aspects of microbial corrosion. The complex role which micro-organisms can play is discussed, along with some of the mechanisms which have been suggested to explain their involvement in the corrosion process. An account is given of the phenomenology of microbial corrosion from a practical engineer's point of view, including the main characteristics and physiological properties of the more. frequently encountered groups of organisms concerned with both anaerobic and aerobic corrosion phenomena, and simple testing procedures for the prediction and diagnosis of problems. The mechanism by which sulphides and elemental sulphur are involved in the corrosion process is also outlined.

Methods available for prevention and protection are reviewed. Indications are given of the gaps in our present knowledge of microbial corrosion and suggestions are made for potentially profitable directions of further research.

Chapter 4 reviews recent research on various forms of localized corrosion with emphasis on testing methods and mechanisms. Recent direct evidence is referenced for the chromium depletion theory of intergranular attack (IGA) in stainless steels and the mechanism of resistance to IGA by duplex phase stainless steels is discussed. The chapter concludes with an interpretive correlation of the mechanisms of pitting and crevice corrosion.

Chapter 5 discusses corrosion fatigue in metals and alloys by adopting as a starting point the review by Gilbert in which pre-1956 knowledge was comprehensively summarized. In general, air fatigue data are used as a basis for comparisons in practice although it is well known that laboratory air is certainly not invariably 'inert' with respect to fatigue crack growth. With the advent of fracture mechanics more meaningful comparisons can be made between crack propagation rates in air and *in vacuo*. Most of this latter work has been carried out on aluminium and its alloys reflecting the importance placed upon a knowledge of the crack propagation rates of these materials for aerospace applications. In the majority of these investigations fatigue crack propagation rates *in vacuo* have been observed to be lower than in air at lower stress levels. However, the effects of gaseous environments are worthy of a separate review and Chapter 5 is concerned with the effect of aqueous environments.

Failure of components operating in various environments due to cracking remains a safety and economic problem, despite the amount of effort that has been devoted in recent times to understanding the phenomena of stress corrosion cracking, hydrogen embrittlement, corrosion fatigue and liquid metal embrittlement. In Chapter 6 the features of the various mechanisms proposed for stress corrosion cracking in ductile alloy/aqueous environment systems are briefly reviewed, followed by an analysis of the problem of quantitatively defining these mechanisms, assessing their validity in a given operating situation and applying this knowledge to the formulation of appropriate remedial actions and design criteria. Finally, recent advances in testing techniques and their application to design criteria are briefly discussed.

During the preparation of this book, it was with great regret that news of the death of Mr V. E. Carter (Chapter 2) was received. Vic Carter graced the corrosion scene for many years, working at the British Non-Ferrous Metals Research Association (BNFMRA; now BNF Metals Technology Centre) for 33 years until 1973. He built an international reputation in his field, where his work ranged from an interest in the testing of metallic coatings in the early years to very significant contributions in atmospheric and marine corrosion in the later years. After leaving BNFMRA he continued to make his wide experience available as a consultant and to committees of such organizations as BSI and ISO. His talents therefore continued to be used and it is fitting that some of these were committed to paper in the chapter that he provided for this book.

R. N. PARKINS

CONTENTS

LIST OF CONTRIBUTORS

The late V. E. CARTER

Corrosion and Metal Finishing Consultant, 19 Newsholme Lane, Durkar, Wakefield, Yorkshire WF4 3BD, UK.

J. CONGLETON

Department of Metallurgy and Engineering Materials, The University of Newcastle upon Tyne, Haymarket Lane, Newcastle upon Tyne, NE1 7RU, UK.

I. H. CRAIG

Department of Metallurgy and Engineering Materials, The University of Newcastle upon Tyne, Haymarket Lane, Newcastle upon Tyne, NE1 7RU, UK.

F. P. FORD

General Electric Co., Research and Development Center, PO Box 8, Schenectady, New York 12301, USA.

D. A. JONES

Associate Professor, Mackay School of Mines, Department of Chemical and Metallurgical Engineering, University of Nevada, Reno, Nevada 89557, USA.

F. MANSFELD

Rockwell International, Science Center, 1049 Camino Dos Rios, PO Box 1085, Thousand Oaks, California 91360, USA.

A. K. TILLER

National Corrosion Service, National Physical Laboratory, Teddington, Middlesex TW11 0LW, UK.

Chapter 1

NEW APPROACHES TO ATMOSPHERIC CORROSION RESEARCH USING ELECTROCHEMICAL TECHNIQUES

F. MANSFELD
Rockwell International, Science Center
California, USA

1. INTRODUCTION

As shown by a number of studies, damage caused by corrosion and the cost of corrosion prevention and control play an important part in the economy of industrialized societies. In 1949 about one-half of the cost of corrosion control in the USA was related to atmospheric corrosion. This includes the use of coatings and paints, galvanizing, phosphating, the use of nickel and cadmium for metallic coatings, and the use of stainless steel and other corrosion resistant materials instead of cheaper metals. The damage caused by failure of bridges, airplanes, etc., is, of course, cause for considerable concern. More recently, corrosion of electronic equipment in commercial and military applications has caused considerable problems. While recent corrosion prevention requirements have decreased airframe-related corrosion problems, corrosion failures of electronic systems still are very numerous and of many different varieties.

Atmospheric corrosion is generally considered to be of an electrochemical nature, and Evans,[1] Rozenfeld,[2] Barton,[3] Kaesche[4] among others have attempted to describe the electrochemical reactions occurring in atmospheric corrosion. Progress in understanding the mechanisms of atmospheric corrosion has, however, been rather slow since, besides general problems in the description of reactions and kinetics of electrochemical corrosion, additional complications arise in

1

atmospheric corrosion. Atmospheric corrosion reactions occur because of the existence of a limited amount of electrolyte, the formation, properties and disappearance of which are influenced by a wide variety of factors including the chemical composition of the atmosphere and the nature of the resulting corrosion products. The fact that the electro-chemical reactions occur under very thin layers of electrolyte is probably the reason why only few electrochemical studies of atmospheric corrosion have been reported. The conventional arrangement for studies of corrosion reactions consisting of small electrodes in large volumes of electrolyte has obviously to be modified for studies of reactions under very thin layers. Similar arguments apply to studies of pitting and stress corrosion cracking of structural materials exposed to the atmosphere. Earlier work by Rozenfeld and his co-workers[2,5] describes experimental arrangements for electrochemical studies under thin electrolyte layers and reports polarization curves under layers of NaCl as thin as 70 μm. The very limited number of electrochemical studies of the kinetics and mechanisms of atmospheric corrosion has no relationship to the vast amount of literature available for corrosion studies in bulk electrolytes or for long-term exposure tests.[6,7] A better understanding of the fundamentals of atmospheric corrosion as affected by the nature of the metal, the composition of the atmosphere, and the time of exposure is necessary if improved measures for prevention of corrosion are to be defined and if such complicated corrosion reactions as those which occur in electronic equipment are to be analyzed.

Atmospheric corrosion is a discontinuous process in which corrosion can occur only when electrolyte layers are present in which the anodic and cathodic reactions that lead to corrosion can take part. The total weight loss, M, during a certain exposure period is the sum of the corrosion losses occurring during n separate corrosion periods of the length $t_{w,i}$:[3]

$$M = \sum_{i=1}^{n} r_i \times t_{w,i} \tag{1}$$

where r is the average corrosion rate during the time t_w, the time-of-wetness. One of the most important aspects of atmospheric corrosion research is, therefore, the experimental determination of corrosion rates, r_i, time-of-wetness, $t_{w,i}$, and frequency of corrosion activity, n, during a given time period. Equally important is the establishment of models of the atmospheric corrosion process which can describe the dependence of r, t_w and n on atmospheric corrosion parameters and the properties of (corroded) surfaces. Electrochemical techniques have the important advantages that the atmospheric corrosion behavior can be monitored continuously and that correlations can be established with atmospheric parameters which are recorded at the same location. The author has recently reviewed briefly the approach of several groups that use electrochemical sensors in atmospheric corrosion research.[8-11] Some of the more important results will be discussed here in more detail.

2. APPLICATION OF ELECTROCHEMICAL SENSORS IN OUTDOOR EXPOSURE

Electrochemical sensors have been used in outdoor exposures mainly to determine the time-of-wetness, t_w, and to establish correlations between t_w and atmospheric parameters such as RH (relative humidity), pollutant concentrations, etc. More recently, attempts have also been made by various authors to calculate corrosion rates from the continuous recording of the sensors and to determine the dependence of corrosion rates on atmospheric conditions.

2.1 Time-of-Wetness

Since atmospheric corrosion can occur only when electrolyte is present on a surface, time-of-wetness, t_w, is obviously a very important parameter. A number of attempts have therefore been made to determine and monitor t_w and establish correlations between t_w and atmospheric parameters. An important point in this context is the consideration that t_w depends not only on atmospheric parameters, but also on the chemical nature of the corrosion products on a metal surface that determine the 'critical RH' above which condensation occurs. Bukowiecki,[12] Preston and Sanyal,[13] Mansfeld et al.,[14-16] and others have investigated this aspect. It seems clear from this consideration that a unique 'critical RH for corrosion' cannot exist for a given metal.

The earlier approaches to measurements of t_w are those of Tomashov[17] and Sereda.[18-20] Tomashov[17] used a 'galvanic corrosion battery' that consisted of alternate copper and iron plates and measured the galvanic current by using a sensitive galvanometer or a recording microammeter. This approach also allows, at least in principle, a determination of the corrosion rate of the anode material, as will be discussed below. The approach of Sereda is different insofar as his sensor only determines whether or not moisture is present on the sensor.

Platinum foils are deposited on steel or zinc panels and when moisture is present, a galvanic current flows which is measured as it flows through a large shunt resistor. Time-of-wetness is arbitrarily defined as the time interval during which this potential exceeds 0·2 V. Sereda has suggested using t_w as the time base for the calculation of corrosion rates as weight loss/t_w;[21,22] he has developed this sensor further and has proposed a miniature version for monitoring surface moisture.[23] Gutman[22] has used Sereda's original device and has shown that t_w measured with this sensor corresponds to the time for which RH exceeded 86·5% based on a four year average at exposure sites in Canada. Table 1 shows t_w data obtained by Sereda for a number of exposure sites and a comparison with RH data which were compiled to show the duration of various RH intervals. Variations in t_w were large from year to year and even large from month to month (Table 1). The reasons for these variations were not clear.

The approach taken by Kucera and Mattsson[24] and Kucera and Gullman[25] which was followed by Mansfeld and co-workers[14-16,26] and others is close to that of Tomashov insofar as galvanic cells such as Cu/steel or Cu/Zn are used. Kucera and co-workers are also using

TABLE 1

PERCENTAGE TIME-OF-WETNESS FOR CANADIAN EXPOSURE SITES AS MEASURED BY SEREDA'S SENSOR (from Ref. 21)

	Halifax 1961–1972	Halifax 1962–1968	Ottawa 1961–1970	Saskatoon 1962–1970	Esquimalt 1961–1969
Yearly average	39.1	44.1	35.3	30.3	42.5
SD	4.9	3.9	4.9	4.5	4.7
Max.	48.4	48.9	41.8	39.5	52.8
Min.	30.7	37.3	29.6	25.4	35.7
RH† equivalent 1955–1966	89.0	87.0	80.0	83.0	87.0
Monthly average Dec.	44.4	52.0	51.6	47.5	61.5
SD	11.6	8.5	9.6	18.0	9.5
Max.	72.3	67.4	71.5	74.8	71.5
Min.	29.1	43.0	38.5	26.2	43.8
Monthly average June	29.5	36.2	26.8	21.4	23.2
SD	6.8	7.6	11.7	7.0	7.4
Max.	47.6	50.4	52.2	29.2	29.5
Min.	21.6	24.4	13.7	8.6	7.0

†RH value derived from meteorological records so that duration of humidity above this value corresponds to the measured time-of-wetness.

passive sensors made of only one metal in a two-electrode configuration to which a constant emf of 100 mV is applied. Mansfeld and co-workers use an emf of ± 30 mV, the polarity of which is changed every 50 s. From this measurement, the polarization resistance can also be determined.

In most cases, the atmospheric corrosion monitor (ACM) consisted of 10 plates of 4130 steel or Zn (99·9%) and 10 parts of OFHC Cu, used in an alternating sequence. These plates were separated by mylar spaces (0.006 cm thickness), the plate thickness usually being 0.06 cm. The plates were arranged in a plastic holder and the whole arrangement was cast in an epoxy resin and then mechanically polished on one side to expose the cross-section of the plates. All copper and all zinc or steel plates are connected together; the galvanic current is measured with a zero-resistance ammeter and a logarithmic converter, which is necessary since the galvanic current changes are many orders of magnitude between the 'dry' and the 'wet' state.

A typical example for the output of a Cu/steel ACM is shown in Fig. 1. During the day-time, current flow is usually not observed in the absence of rain. During the evening hours, some current is registered with the higher values being observed shortly before the surface dries out. This phenomenon, which will be discussed in more detail below (Section 3.1), becomes especially evident in a linear current-time plot which shows that

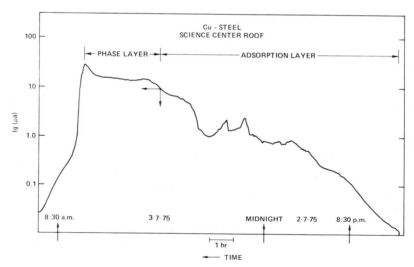

FIG. 1. Current flow for Cu/steel ACM for period of corrosion activity.

the major corrosion activity occurs in the last 4 h before the surfaces dry out, as indicated by the sharp current drop. Further analysis of the results in Fig. 1, and comparison of the log I_g-time curve with data obtained in carefully controlled laboratory experiments, suggests that in the time interval where large fluctuations of the current occur at a level of $I_g \approx 1$ μA adsorbed moisture layers are formed, while at later times where the current is more or less constant at $I_g \approx 10$–15 μA phase layers of electrolyte exist; this analysis follows the suggestion of Mikhailovskii et al.[27] Woelker[28] has shown characteristic current–time traces produced by different types of precipitation (Fig. 2). Rain causes an immediate

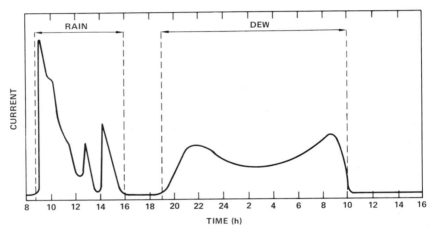

FIG. 2. Current–time behavior produced by different types of precipitation (schematic) (from Ref. 28).

sharp current maximum, and a number of current fluctuations following especially during rain showers. The rate at which surfaces dry after rainfall depends on temperature, RH, and wind velocity and can vary from site to site. The current–time curve caused by dew in general has lower current values and shows much less tendency to show fluctuations.

Figure 1 also serves to illustrate methods for determining t_w. Based on the argument that a corrosion current can flow only when moisture is present, one can conclude that the entire time period of 15 h shown in Fig. 1 corresponds to t_w (see log I_g-time curve). However, the linear I_g-time curve suggests that sufficient electrolyte to cause significant corrosion is present for only about 6 h. This ambiguity which gives a spread between 6 and 15 h in the case of Fig. 1 accounts in part for the large spread in t_w data reported in the literature, a problem discussed further below.

In attempts to correlate t_w with atmospheric parameters, RH, temperature, and t_w are often recorded at the same test site. Figure 3 gives an example for a seven day period from a study at nine test sites in the United States where t_w, RH and temperature, T, were recorded every 5 min 20 s over an 18 month period.[29] All three parameters show the characteristic behavior which has been discussed in more detail elsewhere,[15,16] but RH and I_g (output of a Cu/Zn ACM) change in a very similar manner, with maxima during the night hours, while RH and T change in opposite directions. During most of the day-time, I_g equals the background current of the ACM amplifier which indicates that no corrosion activity occurs, because of the absence of surface electrolyte. The frequency distribution of galvanic current, I_g, RH, and T is given in Fig. 4 for one month at one test site. It can be seen that for this particular case, which is typical for the other eight sites, the surface of the ACM was dry for about 50% of the total time. A shallow current maximum occurs between $I_g = 0.5$ and 3 μA. Relative humidity has a maximum between 85% and 95% where it remains for about 40% of the time.

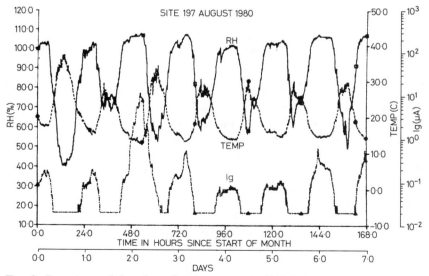

FIG. 3. Experimental data for galvanic current, I_g, RH and temperature T at one site in first week of August 1980.

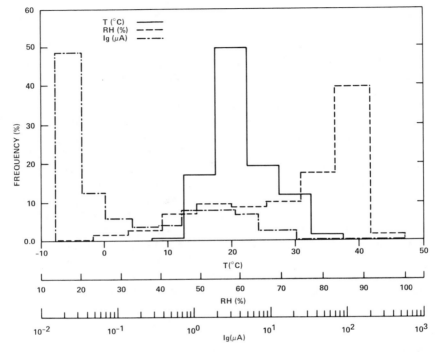

FIG. 4. Frequency distribution of galvanic current, I_g, RH and temperature for one month at one test site.

Another example of the changes of cell current and atmospheric parameters is given in Fig. 5 for exposure in Sweden. These data were reported by Kucera and Mattsson[24] who used an electronic integrator which integrated separately on two counters the amount of current during 'wet' and 'dry' periods. The instrument also records the accumulated time period during which the current exceeds a pre-set value. This time period is considered the time of wetness. While an iron/copper cell was used in Fig. 5, similar data are shown in Fig. 6 for cells of iron, zinc and copper with an imposed emf of 100 mV. Although the current–time curves have similar shapes, during wet periods the current was usually higher for iron cells.

From current–time curves such as those shown in Figs. 1–6 the time-of-wetness can be calculated for given exposure periods. For a three year period in Thousand Oaks, California, the t_w data shown in Fig. 7 have been obtained by using Cu/steel ACMs. For most of the exposure

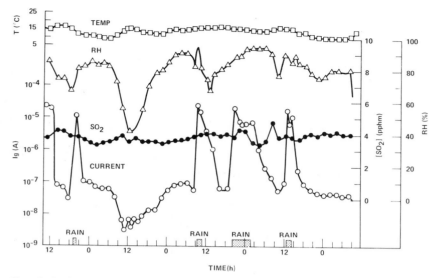

FIG. 5. Fe/Cu sensor output, RH, SO₂ concentration and temperature for a four day period in September 1972 in Stockholm, Sweden (from Ref. 24).

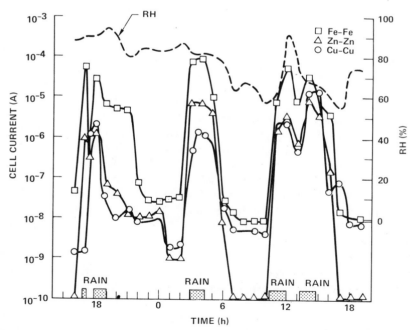

FIG. 6. Output from sensors of three different materials and RH for two days in April 1973 in Stockholm, Sweden (from Ref. 24).

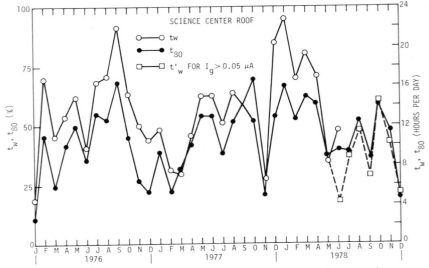

FIG. 7. Time-of-wetness (t_w) and time for RH $\geqslant 80\%$ (t_{80}) for a three year period in Thousand Oaks, California, USA (Cu/steel and Cu/Zn ACMs).

time, t_w was taken as the time for which I_g exceeded the background current of the ACM amplifier (0.02 μA). Starting in June 1978, t_w was redefined as the time for which $I_g \geqq 0.05$ μA (≈ 0.01 μA/cm^2) (t'_w in Fig. 7). Average values of t_w as high as 22.8 h/day (95%) and as low as 4.5 h/day (19%) can occur. For the rather dry year 1977, t_w did not exceed 17 h/day (71%) while for the rainy season between December 1977 and April 1978, t_w exceeded 16 h/day (67%) for five consecutive months. Also shown in Fig. 7 is the time t_{80} for which RH > 80% at the same location. This RH value is considered the 'critical RH' for steel, and Mikhailovskii,[27] Knotkova[30] and others use the time for which RH > 80% and T > 0 °C to calculate t_w. For most of the time period in Fig. 7, $t_w > t_{80}$ which suggests that t_w in its present definition relates to a RH value which is lower than 80%. The t'_w data correspond more closely to t_{80}.

Results obtained by Kucera[25] are given in Fig. 8 for three test sites in Sweden. The time of wetness was 1116, 1190 and 1138 h/year for the sites Vegagatan, Mellbyleden and Floda, respectively, which is about 13% of the total time. When comparing these results with the results in Fig. 7, it has to be recognized that t_w as defined by Kucera is the time for which the sensor output exceeds 1 μA. A similar approach to the results in Fig. 3 would reduce the number of days at which t_w is recorded from seven at

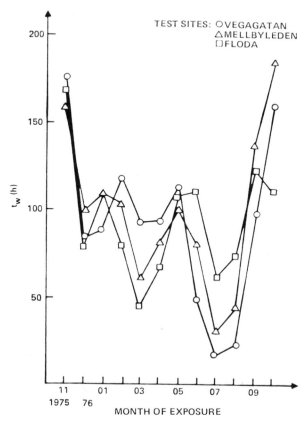

FIG. 8. Time-of-wetness measured with carbon steel sensors (emf $= 100$ mV) for three test sites in Sweden (from Ref. 25).

a limit of $I_g = 0.05$ μA to one at $I_g = 1$ μA. Voigt[31] has used Pt/Fe sensors similar to those of Sereda, stressing the fact that the heat capacity and shape of sensors and corrosion coupons can be made identically in this case. For a four year period in East Germany (Dresden), t_w was between 20% and 80% and the frequency of wetting, n, was between 25 and 98/month. Large fluctuations of t_w for the same months in different years were observed. The monthly averages of t_w and n are shown in Fig. 9 which also includes the time t_{80}^0 for which RH > 80% and $T > 0$ °C. The author concludes that at the beginning of the year a period of large t_w exists which is followed by a period from February to August with an

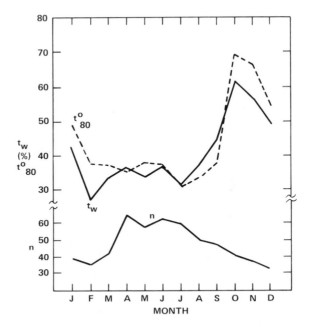

FIG. 9. Monthly average of t_w and frequency n of t_w for 1974–1977 (from Ref. 31).

average of $\bar{t}_w \approx 35\%$. From September on, a period of high corrosion activity is observed which could be a result of increased SO_2 concentration.

A comparison of $\bar{t}_w$ and n in Fig. 9 shows opposite results: periods of high t_w values correspond to fewer periods of wetting. For the example shown for Dresden between 1974 and 1977, it is found that, at the beginning of the year, the wetting cycles are short and their frequency is rising, then starting in August their number decreases, but they last longer.

2.1.1 Time-of-Wetness Based on Corrosion Current Threshold

As discussed above, Kucera et al.[24,25] determine t_w as the time for which a certain sensor current is exceeded since significant corrosion occurs only during this time. Haagenrud[32] tried a similar approach for zinc exposed in Norway, choosing 0.16 μA/cm^2 as the lower limit of the cell current. An analysis of the data obtained over a one year period suggested however, that for zinc, the lower limit should be decreased by about a factor of 10. Figure 10 shows Haagenrud's data which include weight loss of continuously and successively exposed coupons, weight loss calculated from the cell current (zinc cells with an emf of 100 mV), the cumulative time-of-wetness, and monthly values for cell factors. The time dependence of the weight loss for continuously exposed samples, the calculated weight loss, and t_w show similar changes. The cell factor, which is the ratio of the integrated electrochemical data and the weight loss, changes with time and has values between 20% and 51%. The cell factor will be discussed in more detail below.

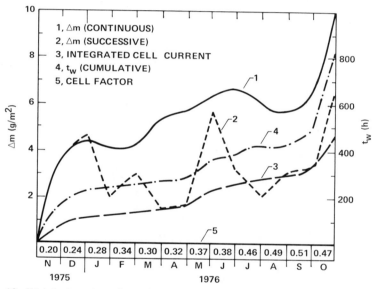

FIG. 10. Weight loss Δm of continuously and successively exposed coupons and weight loss calculated from the integrated cell current, cumulative time-of-wetness, t_w, and cell factor (from Ref. 32).

The choice of a lower limit for t_w is based on Fig. 11 which shows a comparison of electrochemical and atmospheric parameters. The fraction 'wet current/total current' (curve 1) was below 75% for most of the time period in Fig. 11 while a ratio of about 95% was expected since corrosion occurs only when the surface is wet. In addition, t_w (curve 2) was almost identical to the time of precipitation (curve 6) and very much lower than the time for which RH > 90% which, according to Haagerud, is regarded

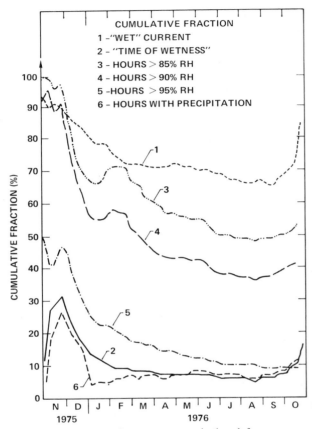

FIG. 11. Cumulative fraction of parameters calculated from sensors and from meteorological data (from Ref. 32).

as the 'real' t_w for zinc. A lowering of the current density level on which t_w is based would move curve 2 in Fig. 11 closer to curves 1, 3 and 4.

In order to avoid some of the uncertainties that can arise when t_w determination is based on a certain level of the sensor output, the author has followed a slightly different approach by determining t_w as the time for which the sensor output exceeds the background current of the sensor amplifier and the time t_{corr} for which the sensor output exceeds 1 μA (0.23 μA/cm^2). In this case, t_w corresponds to the time for which enough condensed moisture is present to allow electrochemical activity, while t_{corr} corresponds to the time for which significant corrosion occurs. The ratio

$\alpha = t_{corr}/t_w$ indicates the fraction of t_w for which significant corrosion occurs. The use of three parameters (t_w, t_{corr} and α) should allow a better characterization of the time dependence of the corrosion activity at a given test site.

This concept has been applied in a program in which the corrosion behavior of solar collector materials has been evaluated.[29] Some experimental data have been given in Fig. 3 for sensor current I_g, temperature T, and RH. Figure 12 shows the changes of t_w, t_{corr} and α for one test site in February 1980. For the entire month, t_w is very high with an average of 23.5 h/day. Between day 6 and day 11, t_{corr} is below 6 h/day, while between day 13 and day 24, it has an average of 14 h/day. Since t_w is more or less constant, α follows the changes of t_{corr}.

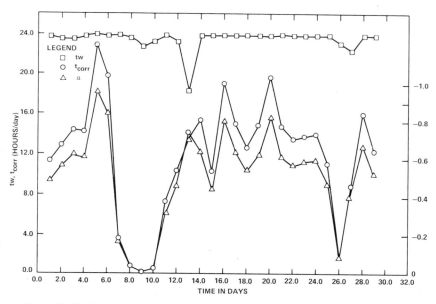

FIG. 12. Daily values of t_w, t_{corr} and α for one test site in February 1980.

The changes of t_w, t_{corr} and α correspond to changes in RH for the same time period. In Fig. 13, the times for which RH exceeds 70% (t_{70}), 80% (t_{80}), and 90% (t_{90}), and the ratio $\beta = t_{90}/t_{80}$ are plotted. The low values of t_{corr} and α between day 6 and day 11 and on day 26 are reflected in the sharp drop of t_{70}, t_{80}, t_{90} and β during this time period. The

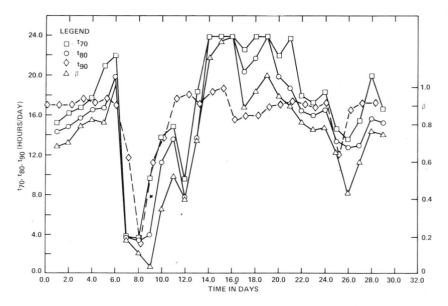

FIG. 13. Daily values of t_{70}, t_{80}, t_{90} and β for same site and same period as Fig. 12.

cumulative integrated charge determined for the Cu/Zn sensor (Fig. 14) shows how the daily corrosion rate changes with time and its relationship to the parameters in Figs. 12 and 13.

Similar analyses have been performed for the entire test period and all test sites, facilitated by the availability of the experimental data on computer tape. The average values of t_w, t_{corr}, α, t_{90}, β, RH, temperature and corrosion loss, Q (Fig. 13), have been used to rank the corrosivity of the nine test sites.[29]

2.4.2 The Time for Existence of Adsorbed and Phase Layers

The preceding discussion has shown the difficulties which seem to exist with the interpretation of electrochemical data in terms of t_w. Mikhailovskii and co-workers[27] do not use a single t_w, but have separated the time for which corrosion activity can occur into the times for which adsorption (τ_{ads}) and phase (τ_{ph}) layers of electrolyte exist. The corrosion loss Δm is expressed in terms of these two parameters:

$$\Delta m = k_{ads}\tau_{ads} + k_{ph}\tau_{ph} \qquad (2)$$

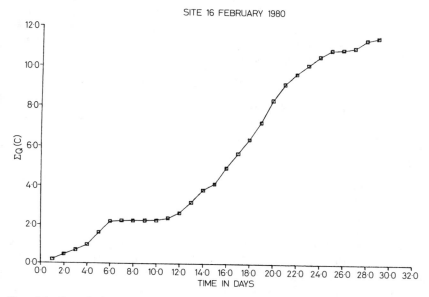

FIG. 14. Cumulative corrosion loss from Cu/Zn ACM for same site and time period as Fig. 12.

where k_{ads} and k_{ph} are the corrosion rates under adsorption and phase moisture films. For the assumption that the corrosion rate of metals is a linear function of the concentration, c_i, of pollutants (gases, salts, etc), Δm is expressed as

$$\Delta m = (k_{ads}^0 + \sum a_i c_i) \tau_{ads} + (k_{ph}^0 + \sum b_i c_i) \tau_{ph} \qquad (3)$$

where a_i and b_i are coefficients which characterize the acceleration of corrosion by pollutants, and k_{ads}^0 and k_{ph}^0 correspond to (pollutant-free) rural atmospheres.

For conditions where phase layers do not occur, such as for exposure in louvered shelters, buildings, etc., $\tau_{ads} \rightarrow 0$ and

$$\Delta m = (k_{ads}^0 + \sum a_i c_i) \, \tau_{ads} \qquad (4)$$

Equations (2)–(4) have been used initially to calculate the atmospheric corrosion rates of zinc and cadmium[27] with the assumptions that k_{ads}^0, which depends on RH, is the average corrosion loss for RH $\geq 80\%$, that k_{ph}^0 is independent of the phase layer thickness, and that the temperature effect can be neglected for $T > -2\,°C$. The time τ_{ads} is calculated from hydrograph and thermograph records as t_{80}^0 (RH $\geq$ 80%, $T > 0$ °C), while τ_{ph} is calculated from moisture sensing elements, pluviograph and dewmeter recordings. For t_{80}^0, a relationship with the average daily RH$_{av}$ was observed for certain sites in the USSR in the form

$$t_{80}^0 = 0.49 \ RH_{av} - 24.9 \ (h/day) \tag{5}$$

According to the authors[27] this equation provides the most probable τ_{ads} values from average daily and monthly RH values; τ_{ph} was calculated from the total duration of periods of rain, dew, and sleet including the drying periods.

The constants k_{ads}^0, k_{ph}^0, a_i and b_i are determined from experimental weight loss data as the slope of weight loss v. τ_{ads} and τ_{ph} in the absence and presence of pollutants.

Figure 15a shows the variations of τ_{ph} and τ_{ads} for a one year period at a subtropical and a tropical maritime region. For the latter region, both values are higher for the entire period beginning with May; the same is true for the ratio τ_{ph}/τ_{ads} (Fig. 15b). For the tropical maritime region, this ratio exceeds 0.7 for eight months; for the same time, it is below 0.4 at the subtropical region. The monthly average RH values do not show large variations and stay between 65% and 85% (Fig. 15b). Although the authors state that τ_{ph} can be calculated from RH$_{av}$ (eqn (5)), it remains doubtful whether monthly average RH values are useful for calculating wetting times. Moreover, it is not clear whether the RH$_{av}$ values were obtained at the test sites or from nearby weather stations.

Experimental values of k_{ads}^0 and k_{ph}^0 are given in the paper under discussion.[27] For zinc $k_{ads}^0 = 8.3 \times 10^{-4}$ and $k_{ph}^0 = 4 \times 10^{-3}$ g/m² h, while for cadmium $k_{ads}^0 = 1.2 \times 10^{-3}$ and $K_{ph}^0 = 6.0 \times 10^{-3}$ g/m² h.

In subsequent papers,[36-38] Mikhailovskii et al. discussed in more detail the determination of the constants in eqns (2)–(4) and have given numerical values for regions of typical climates in the USSR. An interesting result is the much higher τ_{ads} determined in louvered shelters, where $t_{ph} \to 0$. A comparison between calculated corrosion rates (eqn (3)) and observed values resulted in general agreement for zinc and cadmium.

2.2 Corrosion Rates

In addition to mechanistic and modelling purposes, the use of electrochemical sensors for continuous monitoring of atmospheric corrosion rates is of prime importance. Mikhailovskii et al.,[27,36,37] Barton,[3] and Knotkova et al.,[30] among others preferred to calculate

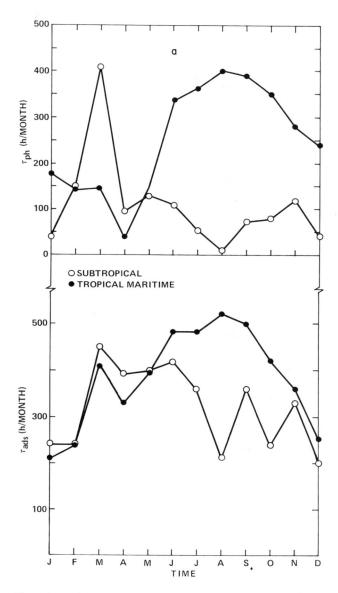

FIG. 15a. Time dependence of τ_{ads} and τ_{ph} for one year at two sites in the USSR (from Ref. 27).

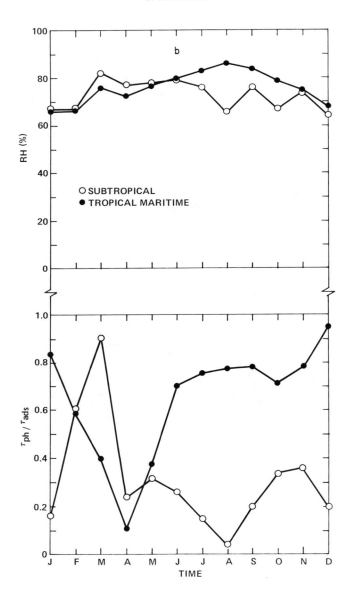

FIG. 15b. Time dependence of τ_{ph}/τ_{ads} and RH for one year at two sites in the USSR (from Ref. 27).

atmospheric corrosion rates from a combination of electrochemical data (mainly t_w, τ_{ph}, τ_{ads}) and meteorological information (RH, t_{80}^0). Since the electrochemical sensor output combines information concerning the corrosion rate of the sensor material, it should be possible, at least in principle, to use the sensor reading to establish a continuous record of the atmospheric corrosion behavior. This makes it possible to examine the corrosion behavior over short time periods (hours, days), when correlations with atmospheric parameters (RH, temperature, pollution) are evaluated, or over longer periods (months, years).

Before the experimental results that have been reported in the literature are discussed, it is necessary to discuss the determination of corrosion rates for different types of sensors.

Experimentally, recording of sensor data is easiest for galvanic cells since no external polarization is necessary. The galvanic current, I_g, is measured by using a zero resistance ammeter (ZRA) and a logarithmic converter.[14] For a cathodic reaction under diffusion control, as would be expected in neutral, aerated solutions, the galvanic current density (cd), i_g^A, for the anode is proportional to the limiting cd for oxygen reduction, i_L:[15]

$$i_g^A = i_L \frac{A^C}{A^A} \tag{6}$$

where A^C/A^A is the area ratio of cathode to anode, and $i_g^A = I_g/A^A$. Since for this case the limiting cd, i_L, equals the corrosion cd, i_{corr}^A, of the anode without coupling to the cathode:

$$i_L = i_{corr}^A \tag{7}$$

i_g^A is a measure of the corrosion cd of the anode:

$$i_g^A = i_{corr}^A \frac{A^C}{A^A} \tag{8}$$

For equal areas $i_g^A = i_{corr}^A$. For this reason, the galvanic current, I_g, can be used to calculate the corrosion rate of the single anode material despite the fact that the anode is polarized in the galvanic couple.

For sensors which operate under an applied emf, the corrosion cd can be calculated, provided the emf is small enough to ensure the validity of the polarization resistance concept.[33] Mansfeld et al. have used steel/steel and zinc/zinc sensors to which an emf of ± 30 mV was applied, the polarity of which was changed every 50 s.[26] In this case

$$i_{corr} = \frac{B}{R_p} = \frac{2b_a \times I_{30}}{2.303\Delta E \times A} = \frac{kI_{30}}{A} \tag{9}$$

where

$$B = \frac{b_a}{2.303}$$

In eqn (9) it is assumed that the cathodic reaction is under diffusion control ($b_c \rightarrow \infty$). A is the area of one of the two electrodes. For an anodic Tafel slope of $b_a = 60$ mV and $\Delta E = 30$ mV, $k \approx 1.75$. The corrosion cd, i_{corr}, is proportional to the measured current, I_{30}. The anodic Tafel slope could change with time and environmental conditions; ΔE could be affected by uncompensated ohmic drop.

For larger polarization, e.g. 100 mV as used by Kucera,[24,25] Haagenrud,[32] and others, eqn (9) does not apply. The corrosion cd, i_{corr}, cannot be calculated without knowledge of the anodic (b_a) and cathodic (b_c) Tafel slopes. Assuming Tafel behavior, the measured current, i_m, equals 2.55 i_{corr} for $b_a = b_c = 120$ mV, while for $b_a = 40$ mV and $b_c = 120$ mV, $i_m = 4.2 i_{corr}$. For diffusion control ($b_c \rightarrow \infty$), the ratio i_m/i_{corr} is even larger. It has been observed in outdoor exposure, however, that the measured current underestimates to a large extent the corrosion rate determined by weight loss.[24,25,32] Figures 16 and 17 show examples from Kucera's work.[25] The output of the sensors (carbon steel and zinc) has been integrated by using an electronic integration to provide monthly weight loss data for steel (Fig. 16) and zinc (Fig. 17). These results illustrate very well the discontinuous nature of atmospheric corrosion. For both materials, a maximum occurs in February 1976 and a minimum in July and August 1976. Figures 18 and 19 show the cumulative weight loss obtained by integration of the sensor current for carbon and weathering steel (Fig. 18) in outdoor exposure and for two Al–Zn–Mg alloys for exposure in a climate chamber with 0.5 ppm SO_2 (Fig. 19). In both cases, the differences between the two materials can be seen to correspond to the known corrosion behavior of these materials in long-term exposure. These results, which were obtained in rather short time periods, contradict the suggestion by Friehe and Schwenk[34] that electrochemical sensors cannot be used for determination of the corrosion behavior of steels and other materials in short-term exposure.

Figure 18 also shows the weight loss of steel and weathering steel panels exposed at the same test site. While the general trends are very

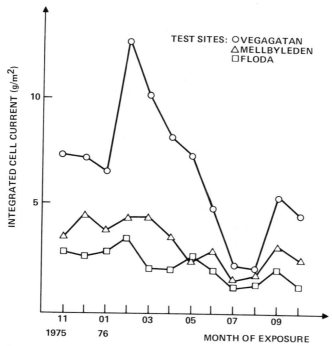

FIG. 16. Integrated cell current from carbon steel sensors expressed as weight loss at three sites (from Ref. 25).

similar to those of the integrated cell current, the weight loss is much higher. Figure 20 shows the relation between corrosion loss as determined by weight loss (Δm) and by the electrochemical techniques (Q) for monthly data obtained by Kucera for carbon steel. The empirical relationship between Δm and Q obtained by least-square analysis shows again that the electrochemical data underestimate the true corrosion rate. The equation in Fig. 20 implies that a weight loss of 15 g/m² can occur without a sensor response ($Q = 0$).

Examples obtained by the author with Cu/Zn ACMs are shown in Fig. 14 for one site and in Fig. 21 for seven sites for one month. The data in Fig. 14 can be compared with the plots of t_w, t_{corr}, and α in Fig. 12 and of t_{70}, t_{80}, t_{90}, and β in Fig. 13. The time periods of low corrosion activity in Fig. 14 correspond to low values of the other parameters in Figs 12 and 13. From the data in Fig. 21, which were recorded in October 1980, one can see longer periods of low corrosion activity and

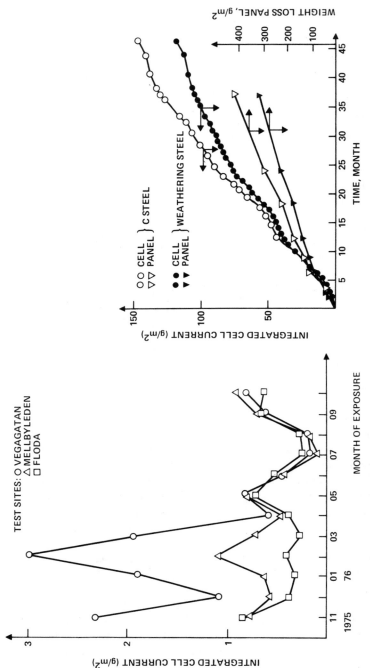

FIG. 18. Integrated cell current from carbon steel and weathering steel sensors converted to weight loss, and weight loss of panels of the same material for time period from June 1975 to August 1979 in Stockholm, Sweden (from Ref. 25).

FIG. 17. Integrated cell current from five sensors expressed as weight loss at three test sites (Ref. 25).

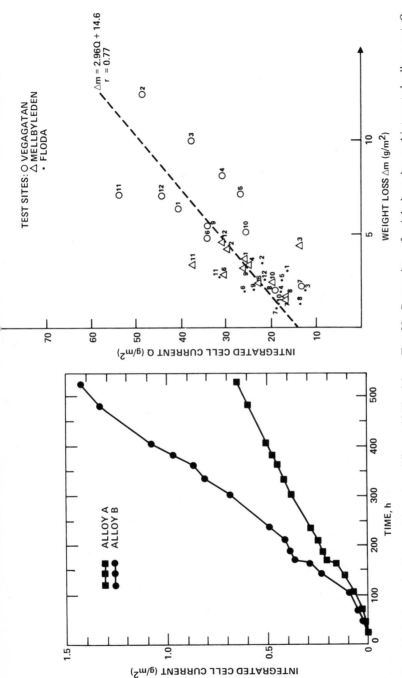

FIG. 19. Integrated cell current from two different Al–Zn–Mg sensors converted to weight loss for exposure to climate chamber (from Ref. 25).

FIG. 20. Comparison of weight loss, Δm, and integrated cell current, Q, expressed as weight loss for time period between November 1975 and October 1976 (numbers indicate the months of exposure) (from Ref. 25).

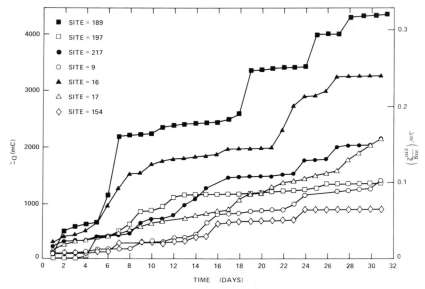

FIG. 21. Cumulative corrosion loss calculated from Cu/Zn ACM data for seven
sites in October 1980.

shorter times of more or less pronounced increases of corrosion rates.
These increases can be clearly seen for sites #189 and #16, while for sites
#9, #154 and #197 a more or less gradual increase of the corrosion loss
occurs. For sites #16 and #189, a weight loss of 0.25 and 0.35 mg/cm²
can be calculated for zinc, which corresponds to an average corrosion
rate of 5 μm/year. Weight loss data from panels exposed at the same sites
were not available for this case.

Figure 22 gives a comparison of the corrosion behavior at two test
sites for the time between August 1979 and October 1980.[29] The average
daily corrosion loss, Q, for each month, as determined by integration of
the cell current and the cumulative corrosion loss ΣQ, shows different
behavior at the two sites. At site #189, wide fluctuations of the corrosion
rate are observed; at site #154, the corrosion rate is almost constant.
Also shown in Fig. 22 are the ratios $\alpha = t_{corr}/t_w$ and $\beta = t_{90}/t_{80}$ which were
discussed above. The limit for t_{corr} was chosen as 0.2 μA/cm² or 3
μm/year for zinc. For the time for which large fluctuations of Q occurred
at site #189, α was about 0.6; at site #154, α was close to 0.4 for most of
the exposure time. The average value of $\beta = 0.84$ at site #189 is very high,
which might be a result of lack of proper calibration of the RH sensor.

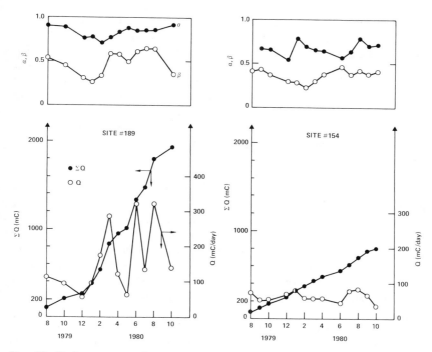

Fig. 22. Daily averages of corrosion loss, Q, and cumulative corrosion loss, ΣQ calculated from Cu/Zn ACMs, and the parameters α and β for two test sites.

This discussion of attempts to use electrochemical sensors to determine atmospheric corrosion rates has shown that the electrochemical data reflect the same trends as the weight loss measurements and give useful information concerning seasonal changes of corrosion rates. However, the discrepancies observed by Kucera et al.[24,25] and Haagenrud[32] between electrochemical and weight loss are a cause for concern. Figure 10 showed 'cell factors' calculated by Haagenrud for zinc as the ratio of electrochemical and weight loss data which change from month to month.[32] Kucera, in analyzing the cell factors for Cu, Zn and steel, found these cell factors to be site dependent.[25] For Cu, values between 0.16 and 0.71 were determined for three test sites; for zinc, the range was 0.14 to 0.48; and for carbon steel, between 0.06 and 0.13. For Al–Zn–Mg alloys, the cell factor determined in 0.1% NaCl spray was between 0.16 and 0.23, while for exposure in a climate chamber with 0.5 ppm SO_2, it was 0.08. For changes in sensor design, Kucera[25] found a decrease of the cell factor with increasing insulator thickness and an increase with increasing emf. As mentioned above, the current measured at an emf of 100 mV should be larger than i_{corr}. The observation of significantly lower values suggests that a number of factors must be involved which change the polarization behavior of these sensors from that expected based on simple electrochemical kinetics. Mansfeld et al.[9,10,26,35] have conducted detailed laboratory studies of the factors which could lead to cell factors of less than 1.0. This problem is discussed in some detail below.

2.3 Correlations with Atmospheric Parameters

A major advantage of the use of electrochemical sensors in outdoor exposure is the possibility of obtaining a continuous record of the corrosion behavior which can be compared with time-dependent atmospheric parameters such as RH, temperature, concentration of pollutants, etc., recorded at the same site. Such a comparison is, in general, not possible with weight loss data in short time intervals, which is probably one of the reasons why development of a better understanding of atmospheric corrosion mechanisms has been so slow.

Gutman and Sereda[21] developed empirical equations for prediction of corrosion rates of steel, copper, and zinc, based on the measurement of atmospheric parameters which included the time-of-wetness t_w, and suggested that corrosion rates should be expressed on the basis of t_w rather than total exposure time. Barton and co-workers[3] have suggested the relationship

$$r = a(t_{80}^0)^n [C_i]^m \tag{10}$$

where r is the steady state corrosion rate, t_{80}^0 the time for which RH $> 80\%$ and $T > 0\,°C$ (from which t_w is calculated by several groups), and $[C_i]$ is the concentration of pollutant i. For relatively constant total pollutant concentration over longer time periods, the corrosion rate can be expressed as

$$r = b(t_{80}^0)^n \tag{11}$$

Haagenrud[32] has established linear relations between the integrated cell current, Q_{corr}, from passive zinc sensors exposed outdoors in Norway with the time-of-wetness, t_w, in the form of

$$Q_{corr}^d = 0.0063\, t_w^d + 0.015, R = 0.94, V = 12\% \tag{12}$$

and

$$Q^{w}_{corr} = 0.0065\, t^{w}_{cw} + 0.066, R = 0.96, V = 4\%$$

on a daily (Q^{d}_{corr}, t^{d}_{w}) and weekly (Q^{w}_{corr}, t^{w}_{w}) basis. Such correlations have to be expected, since both Q_{corr} and t_{w} were determined from the same experimental data. Correlations were also found with the hours and amount of precipitation, and RH.

Mansfeld and co-workers[29] examined their results for Cu/Zn ACMs which were exposed at different sites in the United States and found good correlations between the daily values of the integrated cell current, Q_{corr}, and the corrosion time, t_{corr} (time for which ACM output exceeds 1 μA), in the form

$$Q_{corr} = m t^{n}_{corr} \tag{13}$$

where n and m are constants. Figure 23 is a log Q_{corr}–log t_{corr} plot for the daily averages of these parameters at one test site in February 1980. The values of the constants m and n have been determined by linear least square analysis. The constant m, which equals the average corrosion loss for $t^{n}_{corr} = 1$ h, can be used to characterize the corrosivity of test sites. The

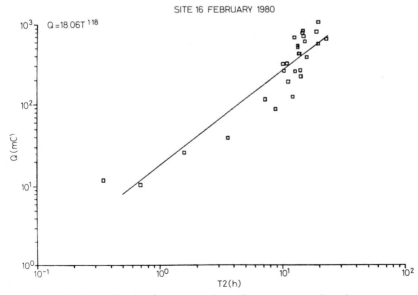

FIG. 23. Dependence of corrosion loss, Q_{corr}, on corrosion time, t_{corr}.

parameter n had values between 0.7 and 0.9 in most cases. The charge Q_{corr} and RH were also highly correlated. However, much more scatter was observed in plots of log Q_{corr} than for plots according to eqn (13). One reason for this behavior is that corrosion occurs only at the higher RH levels.

The approach by Mikhailovskii and co-workers, who calculate atmospheric corrosion rates based on the times τ_{ads} and τ_{ph} during which adsorbed (τ_{ads}) and phase (τ_{ph}) layers are present, has been discussed above. This approach seems to be very successful. In a very recent publication Mikhailovskii et al.[38] discuss a physiochemical model and a corresponding mathematical representation of atmospheric corrosion processes in relation to time, RH, temperature, and the concentration of SO_2 and chlorides in the atmosphere. The probable atmospheric corrosion rates of steel, zinc, copper, aluminum, and a magnesium alloy in outdoor exposure and in a semi-closed atmosphere have been calculated for seven corrosion test sites in Europe. The error of the calculation did not exceed 21%.

3. EVALUATION OF PARAMETERS AFFECTING ELECTROCHEMICAL SENSOR OUTPUT

3.1 The Drying Phenomenon of Thin Electrolyte Layers

The analysis of the time dependence of the ACM signal recorded in outdoor exposure (Fig. 1) has suggested that the main corrosion loss in a 24 h cycle occurs in the short time period when the surface electrolyte layers dry out. In the absence of precipitation in the form of rain, this occurs in the morning hours (Fig. 1) when the temperature increases, RH decreases, and the layers formed by dew become thinner and thinner until all surface electrolyte has disappeared. Despite the large contribution of the drying process to the total corrosion rate, only very few studies of this phenomenon have been performed. Rozenfeld and co-workers[2] have made an important contribution by showing that, in carefully controlled laboratory experiments, corrosion rates are greatly accelerated under thin layers of electrolyte, because of a decrease of the diffusion layer thickness that leads to a higher rate of oxygen reduction, which is the rate controlling step in this diffusion-controlled process. The limiting current density for oxygen reduction, i_L, is inversely proportional to the thickness of the diffusion layer, δ:

$$i_L = \frac{nFDCO_2}{\delta} \tag{14}$$

For constant oxygen concentration, CO_2, and diffusion coefficient of oxygen, D, i_L increases when the diffusion layer becomes thinner. In the atmospheric corrosion process it is possible that CO_2 and D also change during the drying process because of changes of the chemical composition, temperature, etc., of the surface electrolyte. In an attempt to simulate the atmospheric corrosion process, the author has carried out laboratory studies[26] involving electrochemical measurements and a unique weight loss technique under thin electrolyte layers; this is discussed in the following section.

3.1.1 Weight Loss Experiment

In designing the approach for the weight loss measurements, which were carried out for comparison with electrochemical measurements using ACMs, it was important to determine the weight loss under these electrolyte layers the thickness of which changes during the drying process, as observed in outdoor exposure. In order to meet this goal, thin electrolyte layers were placed on samples which were exposed horizontally in test cells with controlled atmospheres and RH $< 100\%$. For comparison, weight loss data were also obtained in bulk electrolyte.[26]

For the thin layer electrolyte studies, flat plates (5 cm × 5 cm) of 4130 steel and zinc (99.9%) were exposed horizontally in glass containers, through which air at constant RH with or without 1 ppm SO_2 was flowing at a rate of 2 litres/min. A 0.5 mm thick layer of electrolyte (0.01 N NaCl, Na_2SO_4, HCl, H_2SO_4 or deionized water) was placed on the samples, the unexposed side of which was coated with an organic coating. Experiments were carried out at RH $= 30\%$, 45%, 60% and 75%. For comparison, weight loss data were also obtained by total immersion in bulk electrolyte for 4 h, which for steel is the time for drying at RH $= 60\%$. The weight loss was determined after removal of the corrosion products, taking into account a correction factor for blank samples. Figures 24 and 25 show the results in air and air $+ SO_2$ for steel and zinc in four different electrolytes.

Increasing RH increases the weight loss for steel in air and in air $+ SO_2$ (Fig. 24), most likely due to the longer time period during which corrosion can occur. An accelerating effect of SO_2 becomes evident at the higher RH values. A comparison of the thin layer and bulk data shows that, particularly in neutral solutions, corrosion is much more severe under thin layer conditions. The results obtained with additions of $NaNO_2$ and the vapor phase inhibitor dichan (dicyclohexylamine nitrite) show that the inhibitors are very effective in NaCl, but provide hardly any protection in Na_2SO_4.

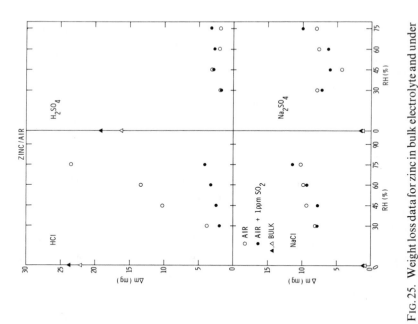

FIG. 25. Weight loss data for zinc in bulk electrolyte and under thin electrolyte layers.

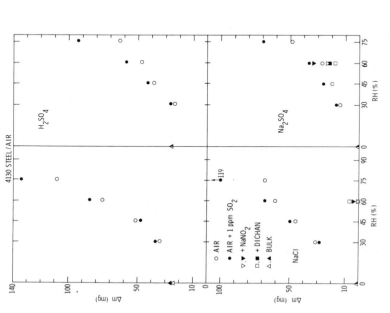

FIG. 24. Weight loss data for 4130 steel in bulk electrolyte and under thin electrolyte layers.

For zinc (Fig. 25), no effect of SO_2 was observed for NaCl and Na_2SO_4. For H_2SO_4, surprisingly, the weight loss Δm was lower than in the neutral solutions and was independent of RH. In HCl, a strong effect of RH was found in air; in air $+ SO_2$, Δm was much lower than in air and was similar to that obtained in H_2SO_4. In NaCl and Na_2SO_4, corrosion rates were much higher under thin layers than in bulk solutions; for HCl and H_2SO_4, corrosion rates were much higher under thin layers than in bulk solutions; for HCl and H_2SO_4, on the other hand, corrosion rates were much higher in the bulk electrolyte.

Some experiments were carried out in argon in order to obtain more information about the mechanism by which SO_2 affects the corrosion process. Much lower weight loss than in air was observed for steel (Fig. 26), which seemed to be independent of RH in the absence of SO_2. Additions of 1 ppm SO_2 increased the weight loss with increasing RH. However, even the highest Δm values are much lower than those observed in air, which suggests that SO_2 does not act as a very effective depolarizer as suggested by Rozenfeld.[5] For zinc, Δm was very low in argon, and no definite effects of RH and SO_2 were observed (Fig. 27).

The effect of electrolyte concentration on corrosion rates under thin layers of electrolyte was investigated in Na_2SO_4 at RH $= 60\%$ (Fig. 28). For both 4130 steel and zinc, an approximately linear increase of Δm with log c was observed.

Experiments were also carried out in deionized water to eliminate the effects of anions on the action of SO_2. The effects of SO_2 were different for steel and zinc (Fig. 29): for steel, Δm increased with increasing RH, and SO_2 accelerated corrosion, while for zinc, RH also accelerated corrosion, but SO_2 had an inhibiting effect.

The corrosion rates calculated from the weight loss data are very high: $\Delta m = 50$ mg for steel corresponds to 1160 mdd or 5.4 mm/year, while $\Delta m = 10$ mg for zinc corresponds to 230 mdd or 1.2 mm/year assuming a corrosion period of 4 h. With respect to atmospheric corrosion conditions, one has to consider that the surfaces are completely wetted only for a fraction of the total exposure time, and that periods of high corrosion activity occur only for very short time periods, as shown in Fig. 1. The electrolyte, most likely, is also less concentrated than 10^{-2} N, which would lead to lower corrosion rates according to the results of Fig. 28.

3.1.2 Electrochemical Experiments

The electrochemical experiments were conducted in the same manner as the weight loss data.[26] Two types of ACMs were used: the active type,

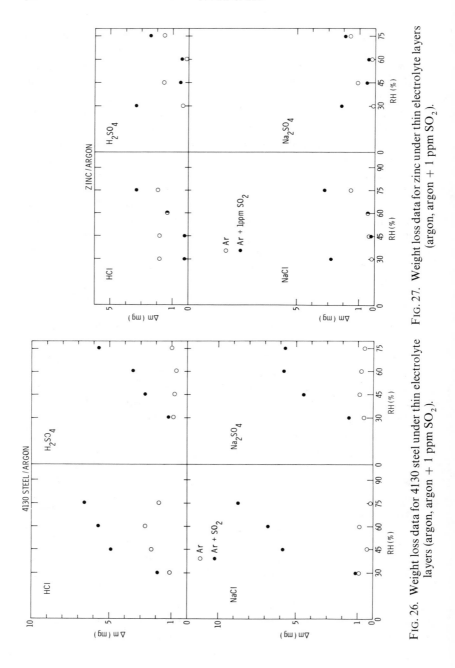

FIG. 27. Weight loss data for zinc under thin electrolyte layers (argon, argon + 1 ppm SO₂).

FIG. 26. Weight loss data for 4130 steel under thin electrolyte layers (argon, argon + 1 ppm SO₂).

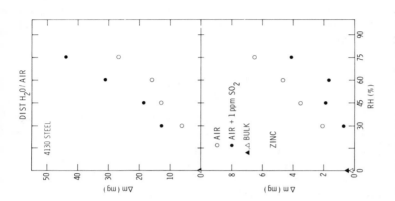

FIG. 29. Weight loss data for 4130 steel and zinc under thin layers of deionized water.

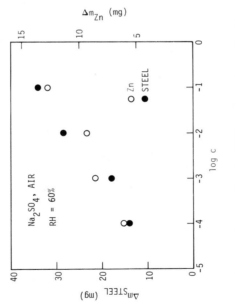

FIG. 28. Weight loss data for 4130 steel and zinc under thin layers as a function of Na_2SO_4 concentration.

which is also being used for outdoor exposure (Cu/steel, Cu/Zn), and the passive type, which consists of plates of one metal (steel or zinc) arranged in a way that a two-electrode system is established. An emf $\Delta E = 30$ mV was applied to these cells, with polarity reversal occurring every 50 s to avoid irreversible changes of the two electrodes. The measured current, I_{30}, is related to the corrosion current I_{corr} by eqn (9):

$$I_{corr} = \frac{B}{R_p} = \frac{2BI_{30}}{\Delta E} = kI_{30}$$

where R_p is the polarization resistance and B is a function of Tafel slopes, having values between $B = 6.5$ and 53 mV for theoretical Tafel slopes,[33] which for $\Delta E = 30$ mV leads to an average of $k = 0.43$ to 3.47.

The ACMs were polished with 600 SiC paper and placed in the glass cells, and a 0.5 mm layer of electrolyte was applied. The current was recorded continuously until the surface layer had dried out, as indicated by a sharp drop of the current. The results obtained for Cu/steel and steel ACMs under a layer of 0.01 N Na_2SO_4 as a function of RH are shown in Figs 30 and 31. The current–time curves are characterized by a continuous increase of the current which becomes

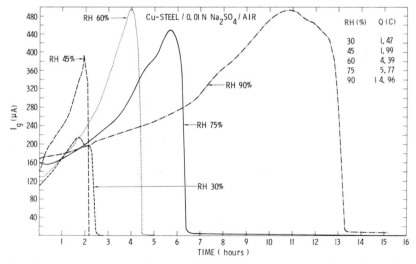

FIG. 30. Galvanic current flow for Cu/steel during drying of a thin 0.01 N Na_2SO_4 layer at different RH values.

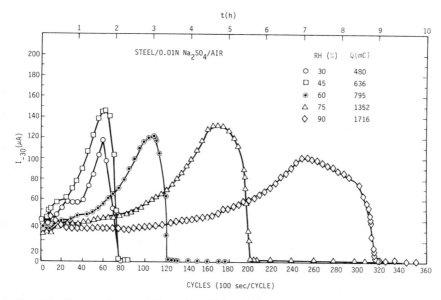

FIG. 31. Current flow at $\Delta E = \pm 30$ mV for 4130 steel during drying of a thin 0.01 N Na_2SO_4 layer at different RH values.

more pronounced when the visible electrolyte layer starts to disappear; this phenomenon is similar to the results obtained in outdoor exposure (see Fig. 1). For both types of ACMs, the time at which the current maximum occurs and the time at which the current decreases to values less than 1 μA both increase with increasing RH, except at RH = 30% and 45% where the main effect of increasing RH is an increase of the current maximum and the total current flow.

For zinc (Fig. 32), differences in the current-time behavior for different RH values are not so pronounced. The time at which the surface dries out is much shorter than for steel, being less than 3 h for RH ≤ 75%. At RH = 75% the current decreases slowly from 3 μA to 0.5 μA over a 3 h period after a large current drop associated with the disappearance of visible electrolyte has occurred. The corrosion loss in this time period as determined by integration of the current–time curve is, however, very small compared to the initial corrosion period. For Cu/Zn, a sharp current maximum is not observed during the drying period (Fig. 33); instead, the current decreases continuously from high values measured

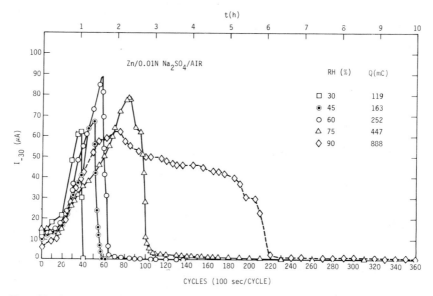

FIG. 32. Current at $\Delta E = \pm 30$ mV for zinc during drying of 0.01 N Na_2SO_4 at different RH values.

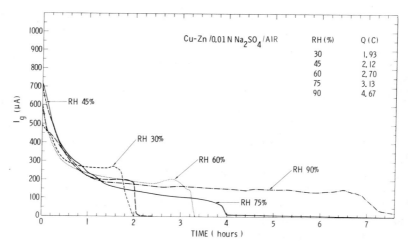

FIG. 33. Galvanic current flow for Cu/zinc during drying of 0.01 N Na_2SO_4 at different RH values.

when the electrolyte was placed on the freshly polished ACM surface, and drops sharply when the surface dries out. The results for RH = 30% and 45% are again rather similar. Drying out occurs after 4 h at RH = 75% and about 7.5 h at RH = 90%.

In order to evaluate the effects of electrolyte concentration, experiments were performed at RH = 60% in which the concentration of Na_2SO_4 was varied from 10^{-1} to 10^{-4} N (Figs 34–37). The characteristic features for both metals shown in Figs 30–33, were preserved, and with decreasing concentration the current flow decreased. For steel and Cu/steel, no clear current maximum was observed at 10^{-4} N. For steel the surface dried out in about 4 h at 10^{-1} N and in 3 h at 10^{-4} N; for Cu/steel the corresponding times were 6 h and 3 h. For zinc and Cu/Zn, drying occurs faster than for steel, as observed in Figs 32 and 33. While the maximum current and the total charge Q increase with RH for both types of sensors, the time for drying seems to be independent of Na_2SO_4 concentration, except for Cu/Zn at $c = 0.1$ N, which is contrary to what was observed for steel and Cu/steel.

The charge Q obtained by graphical integration of the current–time

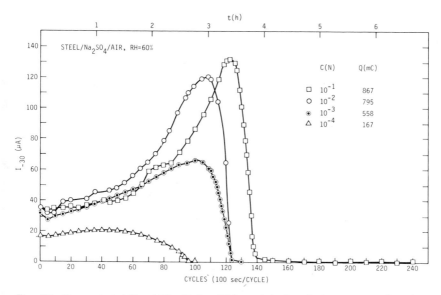

FIG. 34. Current at $\Delta E = \pm 30$ mV for 4130 steel during drying out as a function of Na_2SO_4 concentration.

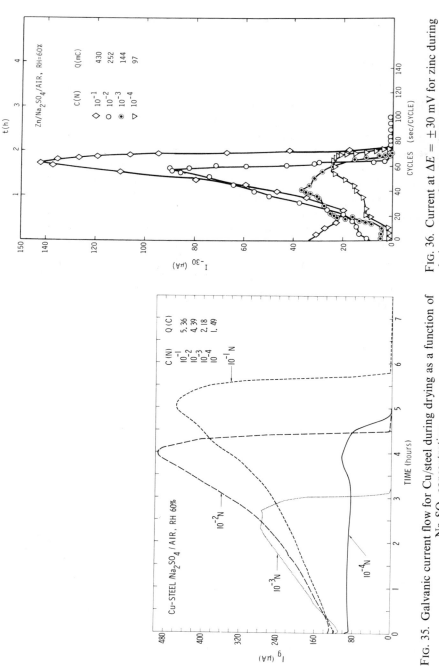

FIG. 36. Current at $\Delta E = \pm 30$ mV for zinc during drying out as a function of Na_2SO_4 concentration.

FIG. 35. Galvanic current flow for Cu/steel during drying as a function of Na_2SO_4 concentration.

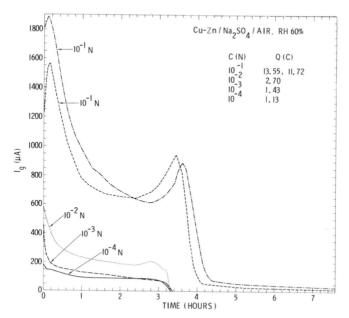

FIG. 37. Galvanic current flow for Cu/zinc during drying as a function of Na_2SO_4 concentration.

curves is plotted in Fig. 38. Higher values of Q are observed for galvanic couples than for one-metal ACMs. However, it should be noted that the charge data for steel and zinc have to be multiplied by the factor k (eqn (9)) to obtain the total charge corresponding to the corrosion loss during the drying out process. The charge data plotted in semi-log form (Fig. 38) as a function of RH suggest a relationship of the form

$$Q = a \cdot e^{b \cdot RH} \qquad (15)$$

Except for Cu/steel, a slope close to unity $(b = 1)$ is observed. This experimental relationship emphasizes again the great importance of RH on corrosion damage.

An example of the importance of the environment on the drying and subsequent wetting process is given in Fig. 39. In this experiment, a Cu/steel ACM was placed horizontally in a glass cell through which air with a constant RH value was flowing at 2 litres/min. The freshly polished ACM was covered with a 0.5 mm thick layer of electrolyte which was allowed to dry at RH = 45%. During this drying process

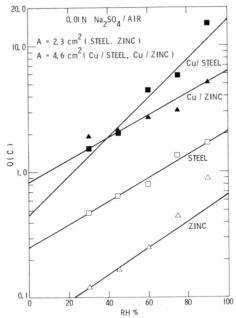

FIG. 38. Integrated current values, Q, as a function of RH for four different ACMs (Figs 34–37).

(indicated by the dashed lines in Fig. 39(a)), the corrosion current increased while the electrolyte layer became thinner. When the surface was dry, the current decreased very sharply to very low values. Current flow was higher in dilute NaCl and Na_2SO_4 than in deionized water, where the time for drying of the surface was longer. After drying, RH was increased to 65%, 80% and 95% for 2 h each, and the output of the ACM was monitored (solid lines in Fig. 39(b)). The current I_g–time behavior in Fig. 39(b)) shows that at equal concentrations, chlorides lead to higher corrosion rates than sulfates. The corrosion products formed in deionized water show very low corrosion currents even at RH = 95%.

3.1.3 Comparison of Weight Loss and Electrochemical Results

Since the weight loss and the electrochemical data were obtained under identical conditions in the laboratory, it is possible to determine the degree to which ACM data provide accurate information concerning corrosion rates. This question becomes even more important when it is considered that Kucera's finding that the current flow for steel sensors

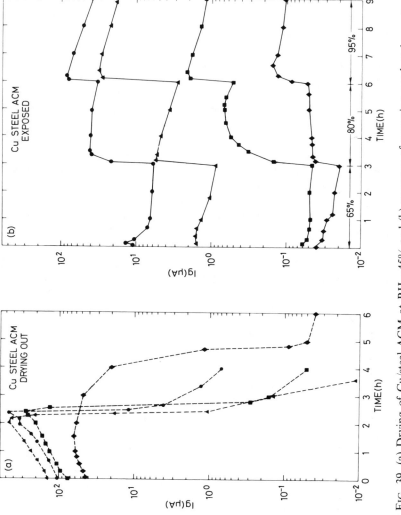

FIG. 39. (a) Drying of Cu/steel ACM at RH = 45% and (b) response of corrosion product layers to different RH levels. ■, 0·001N NaCl; ●, 0·01N NaCl; ▲, 0·01N Na_2SO_4; ◆, DI H_2O.

polarized at constant $\Delta E = 100$ mV accounted for only about 15% of the weight loss observed in outdoor exposure. According to Faraday's law, a weight loss $\Delta m = 1$ mg/cm^2 corresponds to 3.45 C/cm^2 for iron ($z = 2$) and 2.94 C/cm^2 for zinc ($z = 2$). Further discussion of the correlation between weight loss and electrochemical data will be given below when the question of the cell factor is discussed (Section 3.3). A comparison of the weight loss Δm (Figs 24 and 25) and electrochemical data Q per unit area (Figs 30–37) shows that a linear relationship exists for Cu/steel and steel in 0.01 N Na$_2$SO$_4$ at different RH values (Fig. 40(a) and (b)). The results for drying at RH = 60% at Na$_2$SO$_4$ concentrations between 10^{-1} and 10^{-4} N (Fig. 41(a) and (b)) show a much larger change of Δm with Q for Cu/steel than for Cu/Zn (Fig. 41(a)) and an approximately linear behavior for the zinc/zinc ACM (Fig. 41(b)). At constant Na$_2$SO$_4$ concentration, a slope of 0.75 C/mg was observed for Cu/steel; while at constant RH, this slope was 1.09 C/mg. For steel/steel, the slope was 0.28 C(mg in 0.01 N Na$_2$SO$_4$ (Fig. 40b). All these slopes are too low. These results show that the electrochemical data underestimate corrosion rates as determined for weight loss. For Cu/Zn and Zn/Zn, this comparison is more difficult, since the weight loss data showed only a small effect of RH on Δm (Figs 40, 41).

In Table 2, the weight loss data converted into charge data, Q_{WL}, by the use of Faraday's law are compared with the integrated cell current

TABLE 2

COMPARISON OF CHARGE DATA OBTAINED IN 0.01 N Na$_2$SO$_4$/AIR (Q in C/cm^2)

Test condition	Q_{WL}	Q_g	Q_g/Q_{WL}	$Q_{30}^\dagger$	Q_{30}/Q_{WL}
I, RH(%)		A. Steel			
30	1.90	0.320	0.17	0.366	0.19
45	2.75	0.433	0.16	0.483	0.18
60	3.62	0.954	0.26	0.604	0.17
75	6.49	1.254	0.19	1.027	0.16
90	n.d.	3.252	÷	1.306	÷
			0.20		0.18
I, RH(%)		B. Zinc			
30	0.89	0.42	0.47	0.09	0.10
45	0.48	0.46	0.96	0.12	0.25
60	0.87	0.59	0.68	0.19	0.22
75	0.90	0.68	0.76	0.33	0.37
90	n.d.	1.14	—	0.68	÷
			0.62		0.24

$\dagger$for $k = 1.75$ (eqn (9)) ($b_a = 60$ mV, $b_c = \infty$, $\Delta E = 30$ mV)

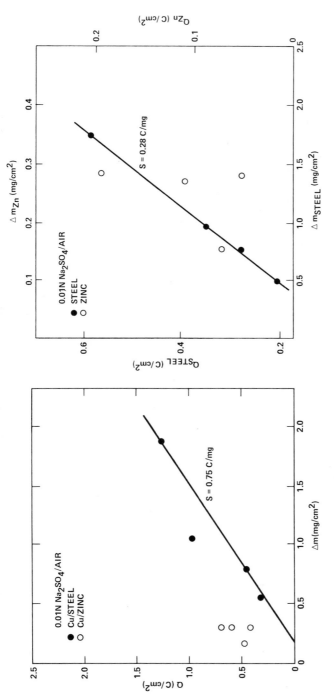

FIG. 40. Correlations between weight loss, Δm, and integrated cell current, Q, per unit area in 0.01 N Na_2SO_4. (a) Cu/steel and Cu/zinc; (b) steel/steel and zinc/zinc.

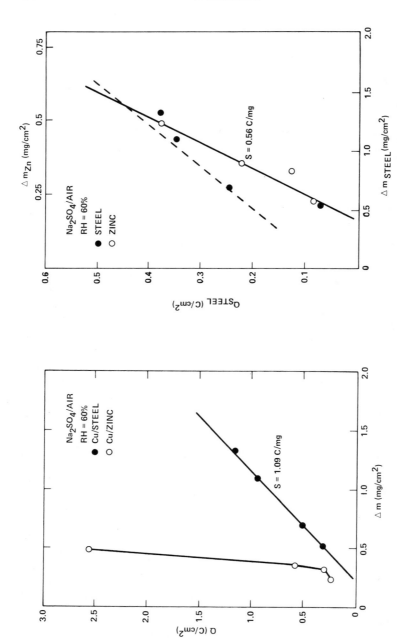

FIG. 41. Correlations between weight loss, Δm, and integrated cell current, Q, per unit area. (a) Cu/steel and Cu/zinc; (b) steel/steel and zinc/zinc.

for galvanic sensors, Q_g, and one-metal sensors, Q_{30}. For steel, remarkable agreement, with a constant ratio $Q_g/Q_{WL} = 0.20$ and $Q_{30}/Q_{WL} = 0.18$, is found; this is close to Kucera's data from exposure tests.[25] For zinc, much larger scatter is observed, with an average ratio of 0.62 for Q_g/Q_{WL} and 0.24 for Q_{30}/Q_{WL}.

3.2 Kinetics of Corrosion Reactions Under Thin Electrolyte Films

Although electrochemical corrosion studies are usually carried out in bulk solutions with small samples which have been carefully cleaned, it is characteristic for atmospheric corrosion that only small amounts of electrolyte are present and only at certain times and that the corrosion product chemistry plays a dominant role. Therefore, in order to evaluate the kinetics of atmospheric corrosion, a different experimental approach has to be used. The approach must take into account the fact that corrosion takes place under thin electrolyte layers, the thickness of which can be time dependent, and that the metal surfaces to be studied are in most cases covered with corrosion products. In addition, the chemical composition of the atmosphere in contact with the electrolyte, which might change because of different levels of gaseous pollutants, must be considered.

A first attempt has been made by the author to develop an experimental approach for the study of atmospheric corrosion under carefully controlled conditions in the laboratory. A study of the literature has shown that only very few thin layer studies which simulate atmospheric corrosion phenomena have been published. Rozenfeld and his coworkers have been very active in this field and an account of their work is given in Rozenfeld's book.[2]

In the earlier studies it became evident that a number of experimental problems have to be solved before electrochemical studies of the kinetics of corrosion reactions can be carried out under thin layers of electrolytes (1000–100 μm). After evaluating the approach of Rozenfeld and coworkers,[2] who used rather complicated arrangements of test, counter and reference electrodes, it was decided that a modified version of the ACM would be best suited for electrochemical corrosion studies under thin layers. The 10 metal plates of an ACM were connected such that the inner two plates served as the reference electrode, the outer four plates were used as the counter electrode, and the four plates between the reference and counter electrodes served as the working electrode. The advantages of this very simple arrangement are that all three electrodes are located in the same plane, which avoids the problems often encountered in other attempts to study corrosion behavior under thin layers.

In the first series of experiments, layers of 0.1 N NaCl or 0.1 N Na_2SO_4 were deposited on steel, Zn, Cu and Al samples in the thicknesses of 1000, 700, 400 and 100 μm and were exposed to laboratory air (RH = 30–50%). After a few minutes at the open circuit potential, a potentiodynamic sweep (2 mV/s) was carried out either in the anodic or cathodic direction. From the anodic curve, the anodic Tafel slope b_a was determined and the corrosion current I_{corr} obtained by extrapolation to the corrosion potential. From the cathodic curve, the limiting current I_L for oxygen reduction was determined. These data were compared with data from metals totally immersed in bulk electrolyte. Figure 42 shows the anodic curves, while Fig. 43 shows the cathodic curves obtained for 4130 steel. Tafel behavior is observed even at 100 μm film thickness; the anodic Tafel slope seems to decrease slightly with film thickness (Fig. 42). These findings are in contrast with those of Rozenfeld et al.,[2] who reported 'anomalous anodic polarization curves'. Their results were probably due to experimental problems related to the fast increase of the current with potential observed in Fig. 42. Also listed in Fig. 42 is the corrosion cd, i_{corr}, which is higher in Na_2SO_4 than in NaCl except at 1000 μm. For comparison, Fig. 44 shows the results obtained for iron in bulk electrolytes. As also observed for Cu, the rate of the anodic reaction is slower in NaCl than in Na_2SO_4.

The cathodic polarization curves in Fig. 43 show an increase of the limiting current, I_L, with decreasing film thickness. The limiting cd, i_L, is inversely proportional to the thickness of the diffusion layer:

$$i_L = nFD \frac{CO_2}{\delta} \tag{14}$$

where $n = 4$, since four electrons are exchanged per molecule of oxygen. If it is assumed that $D = 1.9 \times 10^{-5}$ cm^2/s for the diffusion coefficient of oxygen in 0.1 N NaCl at room temperature and $CO_2 = 2.52 \times 10^{-7}$ mol/cm^3 for oxygen solubility, then:

$$i_L \delta = 1.81 \times 10^{-6} \text{ (A/cm)} \tag{16}$$

Figure 45 plots the limiting cd, i_L, v. the inverse film thickness d for Cu, 4130 steel, and Al 6061 in 0.1 N NaCl and 0.1 N Na_2SO_4. Straight line relationships are observed, with a slope which is close to the one calculated in eqn (16) except for the Al 6061, for which a much lower slope, probably due to the presence of the oxide film is observed. At

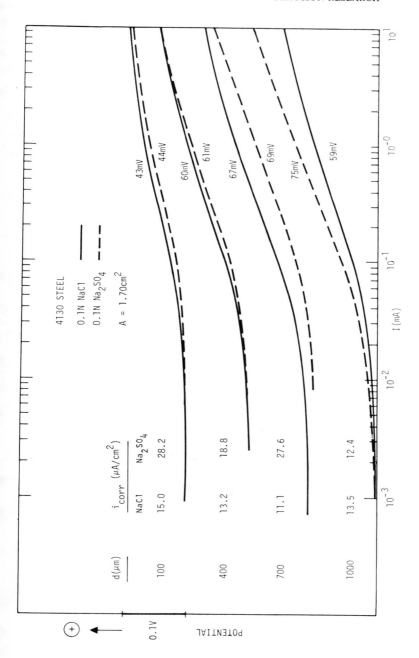

FIG. 42. Anodic potentiodynamic polarization curves for 4130 steel under thin layers of electrolyte.

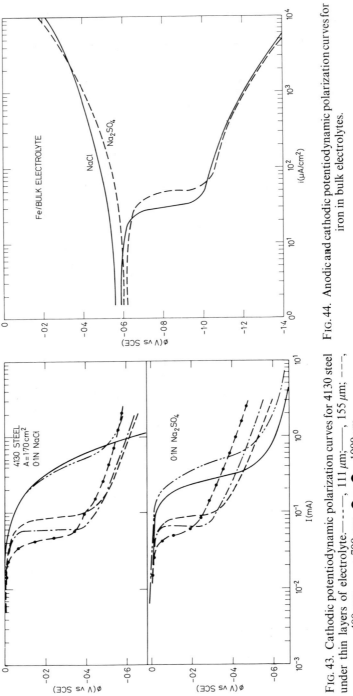

FIG. 43. Cathodic potentiodynamic polarization curves for 4130 steel under thin layers of electrolyte.——, 111 μm;———, 155 μm; – – –, 400 μm;———, 700 μm;– · – ·, 1000 μm.

FIG. 44. Anodic and cathodic potentiodynamic polarization curves for iron in bulk electrolytes.

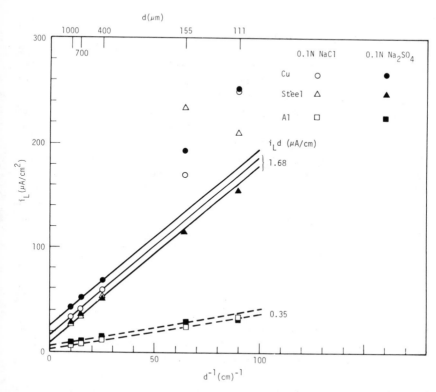

FIG. 45. Plot of limiting cd, i_L, for oxygen diffusion as a function of inverse electrolyte thickness d^{-1} for Cu, 41309 steel and Al 6061.

lower thicknesses ($d = 155$ μm or 100 μm), deviations from the straight lines are observed in Fig. 45, but it is likely that these very thin films evaporate when applied to the metal surface leading to an even thinner film at the time of the measurement. The shape of the cathodic polarization curves in Fig. 43 changes for the lower d values, which makes determination of i_L more difficult. The results in Fig. 45 indicate that for $d < 1000$ μm the diffusion layer thickness corresponds to the electrolyte thickness. With decreasing electrolyte layer thickness, a pronounced increase of the rate of oxygen reduction occurs; this can lead to similar increases of corrosion rates provided that the corrosion reaction remains diffusion controlled even at the lower d values. The drying experiments discussed above have always shown a drastic increase of the galvanic current during the time when the visible electrolyte layer disappeared; this corresponds to a thinning of the film.

Although the polarization curves for steel (Figs 42–44), Cu, Al and Zn showed the general polarization behavior as a function of film thickness, it was also of interest to investigate the corrosion kinetics as a function of time by simulating the drying process and in the presence of pollutants such as SO_2. For these purposes, the recording of polarization curves such as shown in Figs 42–44 is not suitable, since the measuring process alters the nature of the test electrode surface and contaminates the electrolyte with corrosion products and reaction products from the counter electrode. The corrosion cd, i_{corr}, and the Tafel slopes, b_a and b_c, were, therefore, determined from polarization curves obtained within ± 30 mV from the corrosion potential, Φ_{corr}, thereby avoiding excessive polarization. These curves were then analyzed with the computer program CORFIT.[33] Figure 46 shows as an example the inverse polarization resistance R_p^{-1}, the corrosion current I_{corr}, the Tafel slopes and the constant B. Corrosion current and R_p are related by

$$I_{corr} = \frac{b_a b_c}{2.3(b_a + b_c)} \cdot \frac{1}{R_p} = \frac{B}{R_p} \qquad (9)$$

The evaluation with the CORFIT program provides the error of each parameter, which is also shown in Fig. 46. The electrode area was 1.70 cm^2. The results in Fig. 46 show that the cathodic Tafel slope b_c decreases significantly during the time of the experiment, while the electrolyte, which is exposed to laboratory air, dries out. This finding indicates that the corrosion mechanism changes from diffusion control ($b_c \rightarrow \infty$) at the beginning of the experiment to charge transfer control ($b_c = 40$–120 mV) as the electrolyte layer becomes thinner. The anodic Tafel slope b_a, on the other hand, decreases only slightly during the time of the experiment. Similar results were observed in many experiments with steel in NaCl or Na_2SO_4. During the experiment shown in Fig. 46 the corrosion current density decreased slightly from about 50 $\mu A/cm^2$, which is in the range of the values obtained by Tafel extrapolation (Fig. 43). The sharp increase of the limiting current observed during drying in ACM experiments (see Section 3.1.2.) and indirectly in the results of Fig. 46 could not be followed entirely in these measurements, because the potentiostat becomes unstable when the electrolyte layers become too thin.

A comparison of the corrosion cd obtained from 4130 steel from the polarization curves of Fig. 42 and from the results shown in Fig. 46 with

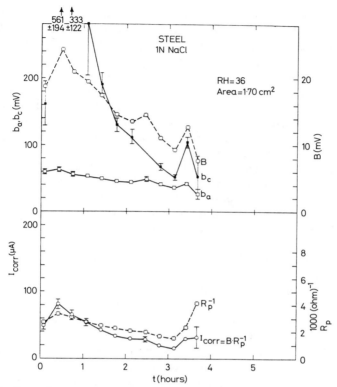

FIG. 46. Corrosion parameters of 4130 steel during drying of a 1 N NaCl electrolyte layer (1100 μm).

the changes of the limiting cd, i_L, for oxygen diffusion and the changes of the cathodic Tafel slope, b_c, during thinning of the electrolyte suggests that while the electrolyte film becomes thinner, the corrosion mechanism changes from diffusion control ($b_c \to \infty$, $i_{corr} = i_L$) to charge transfer control ($b_c \leqslant 120$ mV, $i_{corr} < i_L$). For this case, the anodic and cathodic Tafel lines intersect at a smaller current than the limiting current. While the limiting cd, i_L, increases to 190–230 μA/cm^2 when a film thickness of 100 μm is reached, the corrosion cd, i_{corr}, observed is, in general, below 80 μA/cm^2 even at the time when the electrolyte film is evaporating.

3.3 The Current Efficiency of Electrochemical Sensors (Cell Factor)

In Section 2.2, dealing with results obtained with electrochemical sensors in outdoor exposure, the discrepancy between corrosion rates determined from electrochemical sensors and from weight loss during outdoor exposure was discussed. Kucera et al.[24,25] have determined a 'cell factor', based on weight loss data, which was found to be site dependent. Typical values of the cell factor for iron are about 15%.

It is quite obvious that the question of the (site-dependent) cell factor has to be studied in detail before the electrochemical measurements can be used with confidence and use can be made of the advantages which these measurements provide for monitoring of corrosion rates and/or modelling purposes. A statistically designed experiment has therefore been carried out in the author's laboratory[9,10,35] which addresses the question of the cell factor and the parameters which affect the reproducibility of electrochemical sensor data recorded under simulated atmospheric corrosion conditions. This reproducibility must be established not only for a given sensor as a function of time, but just as importantly, it should be established between sensors. Specifically, the recorded output of a specific sensor should be interpretable in terms of a general corrosion rate and should only depend on the state of the instrument in a known way, so that such a dependence can be accounted for with appropriate calibration procedures.

The reproducibility of the electrochemical measurements has been investigated in tests which involve a drying period under an aqueous solution of 1 mM NaCl followed by either a RH test or a SO_2 test at three different levels. The effect of aging has been studied by outdoor exposure of the sensors for three-month and six-month periods. Fifteen sensors, each of the Cu/steel and steel/steel type have been assembled, using three heats of steel and copper. In this way, for each type of ACM there are three groups of sensors, each prepared from identical materials (same heats of steel and copper, respectively). For the first series of tests, the cleaned ACM surfaces were pre-conditioned by exposure to a 0.5 mm layer of 1 mM NaCl at RH = 45%. After drying, a RH test or a SO_2 test was performed. These tests were run in triplicate. Weight loss samples were also exposed in each test for determination of the cell factor. In the second series of tests the pre-conditioned ACMs were exposed on the Science Center roof for three months. Triplicate RH and SO_2 tests were then performed without cleaning of the ACMs. The third series was carried out in the same manner except for an aging period of six months.

The electrochemical measurements were performed in a glass tube (120 cm long, 15 cm inside diameter) which was closed at both ends, under controlled conditions of relative humidity (RH) and SO_2 concentration.[35] The 15 ACMs were mounted on a support plate and connected to their individual amplifier systems which were directly below the support plate. The ACMs were arranged in statistical distribution of heat (B, C, D) and sensor number (1–5). At both ends of the support plate, three weight loss samples (5 cm × 5 cm) each were positioned. Wet air was fed into the tube from both ends at a flow rate of 4 litres/min by using a plastic tube that had holes drilled into it at intervals which were found to produce a uniform RH distribution in the tube. The desired RH value and SO_2 concentration were produced in a mixing system which has been described earlier. The air exits at both ends of the tube and is fed into a test chamber for monitoring of the flow rate, RH and SO_2 concentration.

Measurements were taken every 50 s and the galvanic current for Cu/steel cells, or the current flow at $\pm$ 30 mV applied emf for steel/steel cells, was recorded sequentially for each of the 15 cells (five times per cell in a 5 s period) by a Hewlett–Packard (HP) multiprogrammer, HP 6940B. Control and timing were provided by an HP computer, 9825S, on which the data were stored for further processing and analysis. After each run, the data were plotted with an HP plotter 9872B for display of the time- and sensor-dependence, followed by integration for the different test intervals, by using a computer program written for this purpose.

From each series of experiments, important information concerning the parameters that affect the output and the reproducibility of ACM data has been obtained. The results obtained in the first test series in which a cell factor has been determined are discussed below.

3.3.1 Effect of Sensor Material and Type

The effects of sensor material and sensor type were especially pro- nounced in the drying tests (first series), as shown in the statistical analysis discussed below. An example of the experimental results ob- tained in the drying test is given in Fig. 47 for the first four sensors on one side of the test tube. All four sensors show very similar behavior. The current–time traces show the typical increase of the current when the electrolyte layers dry out, the drying time, t_{dry}, being between 4.0 to 4.3 h. The RH at the output of the tube stayed very high, despite setting it to 45% at a flow rate of 4 litres/min; this was probably due to the relatively large amount of electrolyte (about 15 ml) on the ACMs and weight loss samples. For the example in Fig. 47(a), RH reached a maximum of 88% after 2.5 h and was at 65% after 6 h when, according to the electrochemi- cal data, the ACM surfaces were dry. This example of RH distribution illustrates the local variations that can occur in outdoor exposure and

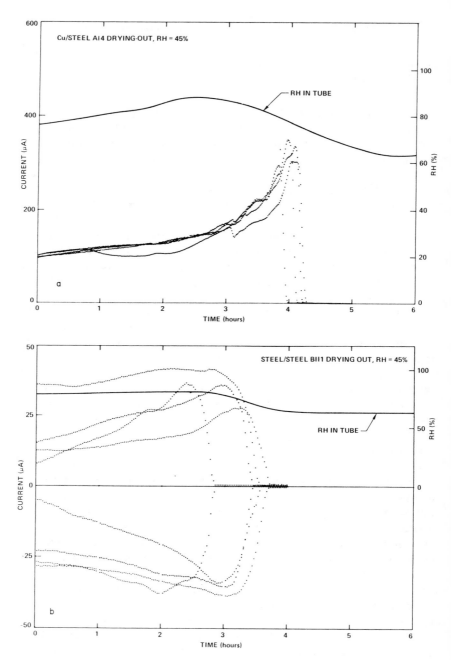

FIG. 47. Experimental results for (a) four Cu/steel ACMs and (b) four steel/steel ACMs drying at RH = 45%.

which would be missed by using average RH values from a nearby weather station.

The current–time behavior during drying of the steel/steel ACMs is shown in Fig. 47(b). A potential of ± 30 mV was applied to the ACMs with polarity reversal after 50 s, which leads to alternating negative and positive currents as shown in Fig. 47(b). In general, an increase of current with time was observed and was more pronounced toward the end of the experiment. Close agreement in the times at which the samples dried out is found for three of the four ACMs in Fig. 47(b), which represent the first four samples on one end of the support plate.

Examples for the response of Cu/steel ACMs to variations of RH or SO_2 concentration are given in Fig. 48. The ACMs were covered with the corrosion products which were formed during the drying period (Fig. 47). When RH was increased from 45% to 65%, an immediate increase of ACM current was observed, followed by a slow decrease (Fig. 48(a)). Similar behavior was found when RH was increased to 80% and to 95%. The spread of the sensor currents reached a factor close to two at the end of the test. Also shown in Fig. 48(a) are the changes of RH observed at the exit site of the test tube. It took about 1 h until RH = 63% was reached, and similar time lags were observed at the other two RH levels. In test cells with only one ACM, the response to RH was usually much faster. The results for the steel/steel ACMs were similar to those shown for Cu/steel.

The current–time curves for four Cu/steel ACMs as a result of variations in the SO_2 concentration are shown in Fig. 48(b). When RH was increased from 45% to 95% without additions of SO_2, there was an initial increase of the current followed by a slow decrease. When 0.2 ppm SO_2 was added to the wet air, this decrease slowed down or stopped in most cases. The SO_2 concentration, which was monitored at the outlet of the test tube, increased slowly and reached 0.18 ppm at the end of 2 h. Other measurements[8] have shown that at lower SO_2 contents it takes several hours before a significant increase of the ACM current is registered. An increase to 1.1 ppm produced an immediate increase of the ACM current showing that there is a definite effect of SO_2 at higher concentrations (and high RH) on corrosion of steel. After aging in outdoor exposure for three or six months, the ACM response to SO_2 was much less pronounced.[35]

The current–time traces have been integrated, by using the HP 9825 S computer, to determine the corrosion loss, Q, during the drying period. These data and the drying times, t_{dry}, are listed in Table 3 for the Cu/steel and in Table 4 for the steel/steel ACMs. The data in Tables 3 and 4 are

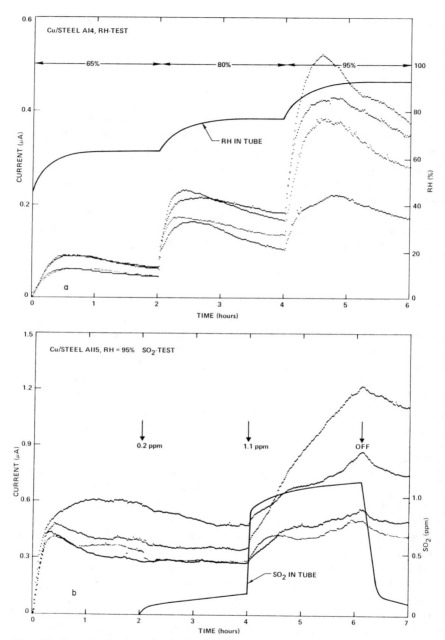

FIG. 48. Experimental results for (a) four Cu/steel ACMs exposed to the RH test, and (b) four Cu/steel ACMs exposed to the SO$_2$ test.

TABLE 3

AVERAGE VALUES OF t_{dry} AND Q FOR DRYING
OF Cu/STEEL ACMs ($A_{steel} = 4.30$ cm^2)

Test number	t_{dry} (h)	Q (mC)
AI1	5.58 ± 0.44	3798 ± 460
AI2	4.56 ± 0.59	3015 ± 298
AI4	4.64 ± 0.49	2302 ± 264
AII3	4.80 ± 0.47	2606 ± 167
AII4	5.70 ± 0.42	3641 ± 246
AI15	4.56 ± 0.31	2333 ± 294
Mean	4.97	2949
SD	0.53	651
V	0.23	353289

TABLE 4

AVERAGE VALUES OF t_{dry} AND Q' FOR DRYING
OF STEEL/STEEL ACMs ($A_{steel} = 2.15$ cm$^2 \times 2$)

Test number	t_{dry} (h)	Q' (mC)
BI1	3.74 ± 0.36	350 ± 60
BI2	3.39 ± 0.31	294 ± 64
BI3	3.17 ± 0.24	265 ± 39
BII1	3.69 ± 0.29	327 ± 42
BII2	3.44 ± 0.29	273 ± 86
BII3	3.40 ± 0.23	295 ± 37
Mean	3.47	301
SD	0.21	32
V	0.037	874

the average values for the 15 ACMs in each test. For the tests for
Cu/steel (Table 3) and steel/steel (Table 4) the mean values, M, the
standard deviation, SD, and the variance, V, have been calculated. The
drying time, t_{dry}, is higher for Cu/steel than for steel/steel ACMs, which is
most likely due to the larger amount of NaCl on the larger surface area
of the Cu/ steel ACMs (8.6 cm^2 v. 4.30 cm^2 for steel/steel). The Q' values
for steel/steel have to be multiplied by the k factor in eqn (9) in order to
obtain the corrosion loss during the drying period. A value of $k = 1.75$
seems to be reasonable. In Tables 5 and 6 the experimental data are
listed again, but this time average values for the five sensors of each heat

TABLE 5

AVERAGE VALUES OF t_{dry} AND Q FOR DRYING OF THREE DIFFERENT HEATS OF Cu/STEEL ACMs

Test no.	Heat B		Heat C		Heat D	
	t_{dry} (h)	Q (mC)	t_{dry} (h)	Q (mC)	t_{dry} (h)	Q (mC)
AI1	5.68 ± 0.58	4267 ± 394	5.62 ± 0.52	3780 ± 608	5.43 ± 0.19	3347 ± 340
AI2	5.12 ± 0.69	3340 ± 620	4.42 ± 0.46	2949 ± 732	4.26 ± 0.35	2755 ± 334
AI4	4.79 ± 0.58	2594 ± 512	4.49 ± 0.43	2080 ± 258	4.65 ± 0.52	2231 ± 375
AII3	4.82 ± 0.61	2787 ± 314	4.69 ± 0.21	2457 ± 147	4.90 ± 0.57	2574 ± 309
AII4	5.67 ± 0.57	3881 ± 393	5.57 ± 0.24	3389 ± 272	4.85 ± 0.43	3653 ± 368
AII5	4.73 ± 0.18	2671 ± 205	4.39 ± 0.41	2197 ± 252	4.57 ± 0.23	2132 ± 322
Mean	5.14	3257	4.86	2809	4.78	2782
SD	0.44	696	0.58	682	0.39	608
V	0.16	403372	0.28	388122	0.13	307993

TABLE 6

AVERAGE VALUES OF t_{dry} AND Q' FOR DRYING OF THREE DIFFERENT HEATS OF STEEL/STEEL ACMs

Test no.	Heat B		Heat C		Heat D	
	t_{dry} (h)	Q' (mC)	t_{dry} (h)	Q' (mC)	t_{dry} (h)	Q' (mC)
BI1	3.95 ± 0.45	395 ± 85	3.57 ± 0.29	328 ± 26	3.71 ± 0.28	327 ± 27
BI2	3.52 ± 0.46	322 ± 95	3.25 ± 0.25	279 ± 48	3.41 ± 0.12	280 ± 40
BI3	3.29 ± 0.17	298 ± 34	3.14 ± 0.17	255 ± 13	3.09 ± 0.34	243 ± 43
BII1	3.86 ± 0.16	365 ± 38	3.45 ± 0.27	293 ± 22	3.76 ± 0.30	303 ± 26
BII2	3.48 ± 0.25	324 ± 35	3.51 ± 0.39	287 ± 46	3.42 ± 0.32	264 ± 19
BII3	3.46 ± 0.20	321 ± 36	3.23 ± 0.22	272 ± 22	3.47 ± 0.24	292 ± 38
Mean	3.59	338	3.39	286	3.48	285
SD	0.255	36	0.23	25	0.24	30
V	0.05	1052	0.04	503	0.05	728

are listed. The t_{dry} data are very similar for all three heats. The corrosion loss, Q, however, shows a strong dependence on metal heat, with heat B having higher Q values than heats C and D which are equal. The weight loss data in Tables 7 and 8 are listed separately for the RH and the SO_2 tests that followed the drying period. However, an analysis of the electrochemical data shows that by far the highest corrosion loss occurs in the drying phase and, as can be seen from the two tables, the weight loss data for the two tests are very similar.

TABLE 7
WEIGHT LOSS DATA FOR 4130 STEEL, RH TESTS (Δm in mg)

Test no.	Heat B	Heat C	Heat D
AI1	25.3	24.0	23.0
	39.7	25.2	26.5
AI2	29.3	29.8	27.6
	23.7	23.3	25.7
AI4	37.0	26.2	24.4
	41.0	20.8	29.4
BI1	29.9	26.7	18.2
	24.7	23.8	25.6
BI2	37.2	28.8	25.8
	22.8	26.7	28.3
BI3	28.2	18.9	23.9
	28.5	25.5	25.3
Mean	30.6	25.0	25.3
SD	6.4	3.1	2.9

TABLE 8
WEIGHT LOSS DATA FOR 4130 STEEL, SO$_2$ TESTS (Δm in mg)

Test no.	Heat B	Heat C	Heat D
AII3	34.4	24.1	28.1
	39.9	31.4	30.7
AII4	37.1	28.7	27.3
	25.7	30.2	25.7
AII5	37.2	25.3	26.6
	33.8	24.7	26.3
BII1	40.2	26.5	23.7
	37.9	20.0	22.6
BII3	31.5	22.4	23.7
	34.6	19.4	26.2
Mean	34.0	25.4	25.3
SD	4.9	3.9	2.8

Heats C and D behave similarly, while heat B corrodes at a higher rate. The average values of t_{dry} and Q are listed for the three heats in Table 9. Although there is no effect of heat on t_{dry}, there is an indication that the electrolyte layer dries out faster on steel/steel than on Cu/steel ACMs. Experiments carried out under identical conditions with ACMs that have different numbers of plates and different plate thicknesses have shown that the total amount of electrolyte on an ACM surface determines t_{dry} for constant external conditions.[35]

The weight loss data have been converted to electrical units in Table 9. A cell factor of about 18% is found for Cu/steel, which is in remarkable agreement with the value of 20% determined for drying under 0.01 N Na_2SO_4 at RH values between 30% and 75% (see Table 2). The cell factor for steel/steel is only about 7%, which is much lower than the average values in Table 2. One reason for this much lower cell factor might be the lower conductivity of 1 mM NaCl, which could lead to a lower effective applied voltage $\Delta E_{eff} = \Delta E - \eta_\Omega$ (eqn (9)), where η_Ω is the uncompensated ohmic drop. These effects are discussed below.

TABLE 9

SUMMARY OF ELECTROCHEMICAL AND WEIGHT LOSS DATA DURING DRYING PERIOD

	Heat B		Heat C		Heat D	
	$t_{dry}(h)$	$Q\,(mC/cm^2)$	$t_{dry}(h)$	$Q\,(mC/cm^2)$	$t_{dry}(h)$	$Q\,(mC/cm^2)$
Cu/steel	4.6	742	4.4	638	4.4	625
Steel/steel	3.1	275	2.9	233	3.0	233
Weight loss (mC/cm^2)						
RH test	4 070		3 310		3 345	
SO_2 test	4 520		3 380		3 345	

Further important information concerning the parameters that can affect the ACM output and its reproducibility, were obtained from an analysis of variance (ANOVA) of the drying data. Since five sensors were fabricated from the material from each heat, it was necessary to analyze sensors as 'nested' factors within a heat. Since only one position could be represented by any one sensor in a given test, it was possible to measure either the effect of relative position or sensor, but not both, for a given analysis configuration. Therefore, two different models were hypothesized and tested. The first model can be expressed as:

$$y_{ijkl} = \mu + \alpha_i + \beta_j + \alpha\beta_{ij} + \gamma_{k(i)} + \beta\gamma_{jk(i)} + \varepsilon_{l(ijkl)} \qquad (17)$$

where

y_{ijkl} = the observation on the jth day using the kth sensor fabricated from the ith heat

μ = an overall mean effect

α_i = the effect of the different heats

β_j = the day-to-day effect

$\alpha\beta_{ij}$ = the interaction effect of heat with day

$\gamma_{k(i)}$ = the effect of different sensors fabricated from the same heat, independently and normally distributed with mean zero and variance $\sigma^2_{\beta\gamma}$

$\beta\gamma_{jk(i)}$ = the interaction effect of day with sensor, the measure of which is considered due only to experimental error, independently and normally distributed with mean zero and variance $\sigma^2_{\beta\gamma}$

$\varepsilon_{l(ijkl)}$ = independent experimental error that is, for every observation, normally distributed with mean zero and common variance; as was explained above, it is not directly measurable.

The hypotheses that were tested were:

$$H_A = \text{all } \alpha_i = 0$$
$$H_B = \text{all } \beta_j = 0$$
$$H_{AB} = \text{all } \alpha\beta_{ij} = 0$$
$$H_C = \text{all } \sigma^2_\gamma = 0$$
$$H_{BC} = \text{all } \sigma^2_{\beta\gamma} = 0$$

In other words, these tests determine if the different levels of each of the main factors (heat, days, and sensors) and interaction factors have the same effect. By using analysis of variances for nested factorial designs, hypotheses H_A and H_B were both rejected with well over 99% confidence, which means that there is a significant heat and day-to-day effect. This day-to-day effect cannot be due to differences in surface preparation, since no differences were found for the sensors of a given heat. It is more likely that day-to-day variations in the RH distribution in the test chamber caused this effect. The hypotheses H_{AB}, H_C and H_{BC} could not be rejected, even for a significance level of 0.25. These results were found for both the copper/steel and steel/steel ACMs.

An important finding was that the sensors were significantly different. The second model was therefore hypothesized in order to test the null hypothesis of no effect due to the relative position of the sensors in the test chamber. The model,

$$y_{ijkl} = \mu + \alpha_i + \beta_j + \alpha\beta_{ij} + \eta_k + \alpha\eta_{ik} + \alpha\beta\eta_{ijk} + \varepsilon_{l(ijk)} \tag{18}$$

is the same as the first model, except that the relative position effect, η_k, is not nested within heat and, hence, its interaction with the other two main effects can be estimated. Specifically, y_{ijkl}, μ, α_i, β_i, $\alpha\beta_{ij}$, and $\varepsilon_{l(ijk)}$ are the same as in the previous model, whereas

η_k = the effect of different positions in the test chamber
$\alpha\eta_{ik}$ = the interaction effect of heat with position (variance $\sigma^2_{\alpha\eta}$)
$\beta\eta_{ik}$ = the interaction effect of day with position
$\alpha\beta\eta_{ijk}$ = the interaction effect of heat, day and position; the measure of this interaction is considered due only to experimental error.

The mean sums of squares for the interaction effects $\beta\gamma_{jk(i)}$ in the first model and $\alpha\beta\eta_{ijk}$ in the second model were found to be nearly equal, thus providing support for the assumption that both are estimates of experimental error. The hypotheses tested were:

H_D: all $\eta_k = 0$
H_{AD}: $\sigma^2_{\alpha\eta} = 0$
H_{BD}: all $\beta\eta_{jk} = 0$

Again by using ANOVA, the hypothesis H_D was rejected with well over 99% confidence, which means that the position of the sensor in the tube affects its output. This is probably due to the presence of the larger amounts of electrolyte on the weight loss samples at the ends of the tube. The interaction effects could not be rejected at a significance level of 0.25. These results were exactly the same for the copper/steel and the steel/steel ACMs, which suggests that the different heats of copper used in the Cu/steel ACMs do not have a significant effect.

The weight loss experiments included the three factors of steel heat, test chamber position (left- or right-hand side of the test chamber), and day. ANOVAs for each of these experiments were in agreement with the electrochemical experiments discussed above in that the difference among heats is statistically significant with more than 99% confidence. The data from these experiments did not provide sufficient evidence to indicate a significant difference between the right- and left-hand positions, or among the days during which the experiments were run.

These results clearly indicate that the heat from which the steel for the fabrication of these sensors is purchased is quite important. In the arrangement used for the experiments, day-to-day variations and chamber location are also significant factors. However, given that the sensors are fabricated with steel from a single heat, they can be expected to provide statistically consistent measurements.

3.3.2 Effect of Electrolyte Conductivity

One reason for the observed low cell factors could be low conductivity of the surface electrolyte layers which are formed in outdoor exposure. For both the galvanic and polarized cells, uncompensated ohmic drop would result in lower cell currents than expected for a given corrosion rate. For the galvanic cells such as Cu/steel, the uncompensated IR drop results in a potential difference between the dissimilar metals and in less than the theoretical (maximum) galvanic current being measured. For the other type of cell, the effective emf is less than the applied ΔE, resulting in a lower current. The author has carried out experiments in the laboratory in order to determine the relative magnitudes of the polarization resistance, R_p, and the uncompensated resistance, R_Ω.[39] The measurements used the ac impedance technique as described elsewhere[40] and a potentiostatic technique with automatic IR drop compensation.[41]

The ac impedance measurements, which were carried over a frequency region of 10 kHz to 25 mHz with an ac signal of 10 mV, not only provide information concerning the magnitude of R_p and R_Ω, but can also be used for mechanistic studies. Tests were performed with Cu/steel and steel/steel ACMs under 0.5 mm layers of 1 mM NaCl and 5 mM Na_2SO_4 in a glass cell at RH = 100% 1 h after application of the electrolyte and after the surface had dried at RH = 40%. The results are shown as Bode plots (Z is the absolute value of the impedance and $\omega = 2\pi f$ is the frequency in rad/s) in Fig. 49 for Cu/steel and Fig. 50 for steel/steel. From the horizontal part of the Bode plot at high frequencies R_Ω can be determined, while R_p is determined from the impedance limit at the lowest frequency. The solution resistance is higher in 1 mM NaCl (about $200\,\Omega\cdot cm^2$) than in 5 mM Na_2SO_4 ($30\,\Omega\cdot cm^2$). However, in both solutions, R_Ω is much lower than R_p. As long as liquid layers are present on the sensor surface, uncompensated ohmic drop should not produce significant error in determining the corrosion current. Further analysis of the impedance data shows deviations from the ideal semicircle in the complex impedance plane[40] which are more pronounced for Cu/steel sensors. The observed depressed semicircles are considered to result from the heterogeneity of the sensor surface.

When the sensor surface became dry, the impedance plot was that for a very large resistor ($R_p > 10^5$ ohm) with a small capacitor, C, in parallel. Since similar values for C ($\approx 5 \times 10^{-9}$ F) and R_p ($\approx 10^7$ ohm) have been measured for dry, polished sensors, it can be concluded that very little corrosion activity occurs at RH = 40% after drying. The observed capacitance is considered to be due to the mylar film used as spacers, while

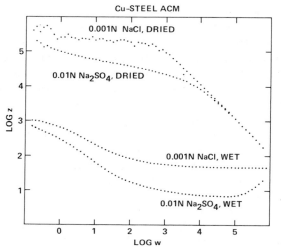

FIG. 49. AC impedance plots for Cu/steel ACMs.

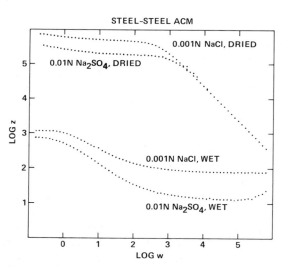

FIG. 50. AC impedance plots for steel/steel ACMs.

the resistance is due to residual electrical conductivity of the corrosion products bridging the individual plates.

The results from the experiments with automatic IR drop compensation are shown in Figs 51 to 54. The interrupter unit PU1 (Meinsberg, East Germany) was used in combination with a PAR potentiostat. For Cu/steel drying at RH$=35\%$ under 0.01 N Na$_2$SO$_4$ (Fig. 51) and

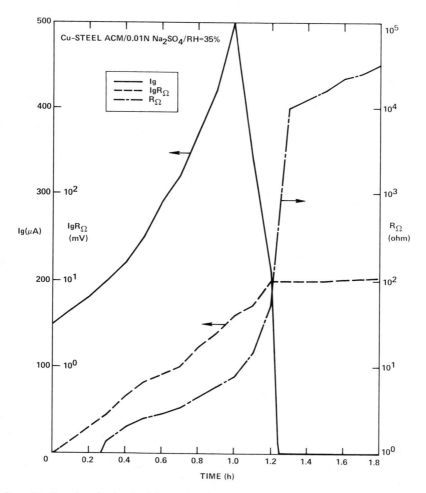

FIG. 51. Results obtained with automatic compensation of ohmic drop (Cu/steel ACM drying under 0.01 N Na$_2$SO$_4$).

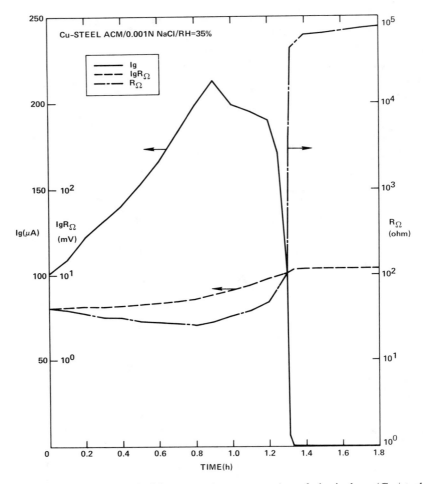

FIG. 52. Results obtained with automatic compensation of ohmic drop (Cu/steel ACM drying under 0.001 N NaCl).

0.001 N NaCl (Fig. 52), the typical current, I_g–time behavior is observed; however, the maximum current is higher than determined before without IR drop compensation (see Section 3.1.2). Also recorded in these experiments is the ohmic drop, $I_g R_\Omega$, which is also shown in Figs 51 and 52; for both solutions, it increased continuously to about 10 mV and did not change much when the liquid electrolyte layer had disappeared. From

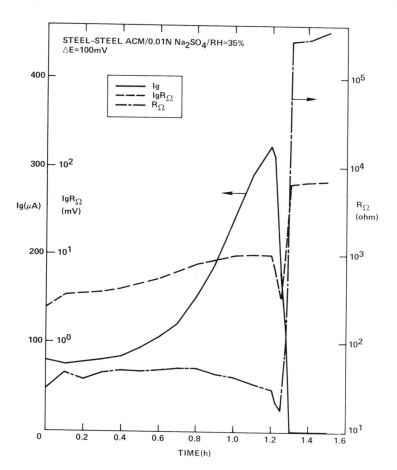

FIG. 53. Results obtained with automatic compensation of ohmic drop (steel/steel ACM drying under 0.01 N Na$_2$SO$_4$).

the recorded values of I_g and I_gR_Ω, the solution resistance R_Ω can be calculated. For both solutions, R_Ω reaches about 100 ohm while the solution evaporates, and exceeds 10^4 ohm when the visible electrolyte layers have disappeared. For steel and zinc sensors with an applied $\Delta E = 100$ mV (Figs 53 and 54) under 0.01 N Na$_2$SO$_4$ at RH $= 35\%$, the current-time curves show typical behavior. The ohmic drop is less than 10 mV as long as liquid layers are present, but reaches about 100 mV

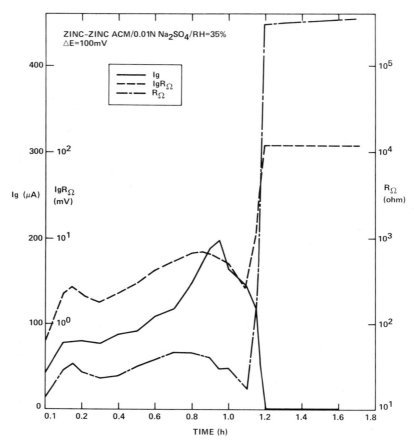

FIG. 54. Results obtained with automatic compensation of ohmic drop (zinc/zinc ACM drying under 0.01 N Na_2SO_4).

when these layers have disappeared. The ohmic resistance, R_Ω, changes from less than 100 ohm to more than 10^5 ohm.

In these experiments, the ohmic drop was always less than 10 mV as long as the cell current was more than 1 μA, which is the lower limit for the time-of-wetness in Kucera's analysis.[24,25] It could therefore be argued that the conductivity of surface layers during periods of active corrosion cannot play a major role in producing the observed low cell factors. However, more experiments using ac impedance or the inter-rupter technique have to be performed during outdoor exposure of the electrochemical sensors in order to determine accurately the conditions occurring during the wetting and drying conditions in real atmospheres.

3.3.3 Effect of Sensor Design

Other factors which could play a role in determining the efficiency of the electrochemical sensors in the measurement of atmospheric corrosion rates could be the type of sensor, such as galvanic couples (active) v. sensor with applied emf (passive) and design of sensor (size, heat capacity, number and thickness of plates, etc.). Voigt[31] has reported results of time-of-wetness, t_w, measurements with different sensors which included the Pt/Fe sensors as suggested by Sereda[18-20] and passive sensors developed at the Akimov Institute in Czechoslovakia (QAY system and QAY system with Fe/Fe sensor). Table 10 shows t_w data for

TABLE 10

EXPERIMENTAL VALUES OF TIME-OF-WETNESS, t_w, IN HOURS (FROM REF. 31)

Month	Pt/Fe	Sensor QAY	QAY with Fe/Fe element	Test duration
4/1977	239 (33.2)	294 (40.8)	412 (57.2)	720
5/1977	213 (28.6)	304 (40.9)	265 (35.6)	744
6/1977	298 (41.4)	320 (44.4)	358 (49.7)	720
7/1977	242 (34.7)	321 (46.1)	355 (51.0)	696
8/1977	376 (46.1)	417 (51.1)	449 (55.0)	816
9/1977	397 (55.1)	415 (57.6)	351 (48.8)	720
12/1977	487 (75.4)	527 (81.6)	395 (61.1)	646
1/1978	209 (27.2)	487 (63.4)	688 (89.6)	768
2/1978	206 (29.6)	432 (62.2)	390 (56.1)	695
3/1978	252 (41.2)	287 (46.9)	367 (60.0)	612
Sum	2 919 (40.5)	3 803 (52.8)	4 030 (56.0)	7 197

Note: Values in brackets are t_w in %.

these three sensor types. The trends for the three sensor types are very similar for most of the time except for January 1978, where the Pt/Fe sensor shows a minimum while for the QAY system with Fe/Fe sensor a maximum was observed. The mean values for the entire period for the two QAY systems differed only by about 65%, but were higher by about 12% to 16% than the Pt/Fe sensor. The Pt/Fe sensor produced t_w values that were very close to the t_{80} values, but considerably larger than the t_{90} values. In discussing these data, Voigt concluded that t_w as a measure of the length of corrosion periods depends on the measurement principle (active v. passive sensors) due to different sensitivities. A passive system that uses an applied emf could produce t_w values that are too high, since processes occurring under very thin electrolyte layers that are not significant for the corrosion process are included. However, the present discussion has shown that the highest corrosion losses occur under the thin layers in the drying period (see Section 3.1). Additional contributions could arise from electrolysis of surface electrolytes due to the (rather arbitrarily chosen) external emf. Voigt also concluded that the Pt/Fe sensors according to Sereda would measure t_w data which are too short due to rupture of the electrolyte film at corners, etc., of the Pt/Fe sensor and breakdown of the galvanic driving force.

This discussion of Voigt's results is included here, since observed differences in t_w measurements with different sensor types point to the possibility of differences in corrosion rates as determined by integration of the current–time curves measured by such sensors. Kucera[25] has evaluated the effects of insulation thickness and impressed emf on t_w and cell factor (Table 11) and found that the sensor current decreases linearly

TABLE 11

INFLUENCE OF CELL CONSTRUCTION AND IMPRESSED VOLTAGE ON CURRENT FROM C STEEL CELLS DURING A ONE MONTH EXPOSURE IN STOCKHOLM (FROM REF. 25)

Insulation thickness (μm)	Impressed voltage (mV)	Time-of-wetness (h)	I Weight loss cell (g/m^2)	II Weight loss panel (g/m^2)	Cell factor I/II
100	100	79.1	2.8	19.8	0.14
200	100	75.0	2.1	19.8	0.10
400	100	63.8	1.4	19.8	0.07
100	100	79.1	2.8	19.8	0.14
100	200	78.4	4.4	19.8	0.22
100	300	81.6	5.6	19.8	0.28

with increasing insulator thickness and increases linearly with emf. The time-of-wetness decreased with increasing insulation thickness and increased with increasing emf. When the plate thickness of carbon steel was increased from 0.5 to 0.9 mm, the cell factor decreased by about 30% and t_w also decreased.

In the experiment conducted by the author[35] which has been described above, four sensors each of the active and passive type, that had different numbers and thicknesses of plates (Table 12), have been used.

TABLE 12

SUMMARY OF DRYING DATA FOR ACM OF
VARYING DESIGNS

ACM	t_{dry} (h)	Q_{corr} (mC/cm^2)
E6	4.91 ± 1.24	225
E3	5.82 ± 0.88	688
E7	5.93 ± 0.78	307
E8	6.06 ± 0.90	309
E1	4.83 ± 0.64	133
E4	4.62 ± 1.02	52
E2	6.15 ± 0.64	152
E5	6.33 ± 1.16	100

Construction of these ACMs was the same as for the standard ACM (Section 2.1); however, a different steel (A366) which is close to a carbon steel, Type 1015, was used. Higher cell factors than for the Type 4130 steel were observed. The average weight loss for drying under a 1 mM NaCl layer of 0.5 mm thickness corresponded to 1190 mC/cm². A cell factor of 58% was found for the sensor E3 which has 10 plates each of steel (0.91 mm thickness) and copper (0.51 mm thickness). For the same number of plates, but half the thickness of steel plates, sensor E6 had a cell factor of only 19%, which is a surprising result since it was expected based on Kucera's results (Table 11), that thinner plates would lead to higher cell factors. Sensors E7 and E8 which have 20 plates each of steel (0.46 mm thickness) and copper (0.25 mm thickness) and differ in the insulator thickness by a factor of two, have cell factors of 28%. In a similar manner, the steel/steel sensor E2 with 20 plates of 0.91 mm thickness has a cell factor of 13%, while sensor E4 with the same number of plates but half the plate thickness has an efficiency of only 4%.

4. SUMMARY

This review of recent applications of electrochemical sensors to atmospheric corrosion research has shown that important information can be gained in outdoor exposure as well as in laboratory experiments in which the influence of the many factors that play a role in atmospheric corrosion reactions can be studied. The possibility of obtaining continuous records of atmospheric corrosion behavior makes it possible to determine the effects of atmospheric parameters recorded at the same site.

Measurements of the time-of-wetness have been reported by a number of groups. This parameter is undoubtedly very important in atmospheric corrosion studies; its interpretation has, however, caused a number of problems which have been discussed in this review. Electrochemical sensors provide a signal whenever sufficient electrolyte to allow current flow is present. However, during most of the 'wet' period, the corrosion rates are very low, which has led some authors to the definition of t_w as the time for which a minimum sensor current is exceeded. Since different minimum current levels are used by different groups for the same material and different minima for different materials, one cannot expect good agreement between results reported in the literature.

Mikhailovskii *et al.* base their calculations of atmospheric corrosion rates on the times for which adsorbed (τ_{ads}) and phase (τ_{ph}) layers are present on the wetted surface. The time τ_{ph} is calculated from electrochemical sensors, while τ_{ads} and the constants needed to calculate corrosion rates are obtained from meteorological data and from weight loss measurements. The author has proposed to determine t_w as the time for which an electrochemical sensor produces a signal and t_{corr} as the time for which a certain corrosion current is exceeded. These two times are considered to be equivalent to τ_{ads} and τ_{ph} used by the Soviet group.

An important aspect of atmospheric corrosion is the determination of the corrosivity of the atmosphere. Here, electrochemical sensors undoubtedly can play an important role while providing numerical values of t_w, t_{corr} and $\alpha = t_{corr}/t_w$ as suggested by the author. Working group 4 of the International Standards Organization's Technical Committee 156 is working on standards that can be used to characterize the corrosivity of test sites.

Much progress has been made in the use of electrochemical techniques for corrosion monitoring and control in industrial applications where sensors are exposed in bulk situations. In atmospheric corrosion conditions, only a beginning has been made in attempts to monitor continuously the corrosion rate of metals at different test sites. In most cases the electrochemical sensors have determined only small fractions of the corrosion rate of metal coupons exposed at the same test sites. In evaluating these results, it has to be remembered that the electrochemical sensors can determine only the corrosion rate of the sensor material. The author has made an attempt to analyze the factors that can affect the response of such sensors to changing atmospheric conditions and to determine the causes of the observed low cell efficiency. Much more research is needed to develop an approach for quantitative corrosion rate measurements under atmospheric corrosion conditions which are much more complicated than those of exposure to bulk electrolytes.

Despite the present problems in determining quantitative corrosion behavior, electrochemical sensors should be very helpful in establishing models for predictions of long-term corrosion behavior. Empirically it has been observed from weight loss data collected over several years that corrosion rates r change with time t according to[42-44]

$$r = kt^{-n}$$

The use of electrochemical sensors in outdoor exposure with statistically designed tests should allow the determination of the dependence of the parameters k and n on atmospheric conditions. The parameter k can be used to characterize the corrosivity of a test site, while the parameter n depends on the corrosion mechanism. By providing reliable values of k and n in short exposure times, the electrochemical techniques would produce major progress for practical applications under atmospheric corrosion conditions.

REFERENCES

1. U. R. Evans, *Nature*, **206**, 180 (1965); *Corr. Sci.*, **9**, 813 (1969); *Corr. Sci.*, **12**, 227 (1972).
2. I. L. Rozenfeld, *Atmospheric corrosion of metals*, NACE, Houston, Texas, 1972.
3. K. Barton, *Schutz gegen atmosphaische Korrosion*, Verlag Chemie, Weinheim/Bergstr, West Germany, 1973.
4. H. Kaesche, *Werkstoffe und Korrosion*, **5**, 379 (1964).
5. I. L. Rozenfeld, *1st Int. Cong. Metallic Corrosion*, London, 1961, Butterworth, p. 243.
6. Metal corrosion in the atmosphere, ASTM STP 435 (1968).
7. W. H. Ailor, ASTM STP 646, 129 (1978).
8. F. Mansfeld, *Proc. 5th European Symp. on Corrosion Inhibition*, Ferrara, Italy, Sept. 1980, pp. 191–216.
9. F. Mansfeld, S. Tsai, S. Jeanjaquet, M. E. Meyer, K. Fertig and C. Ogden, Reproducibility of electrochemical measurements of atmospheric corrosion phenomena. ASTM STP XXX on atmospheric corrosion (in press).
10. F. Mansfeld, Electrochemical methods for atmospheric corrosion studies, *Proc. Symp. on Atmospheric Corrosion*, The Electrochemical Society, Hollywood, Florida, Oct. 1980 (in press).
11. F. Mansfeld, ASTM STP 727, 215 (1981).
12. A. Bukowiecki, Schweiz. Archiv., **23**, 97 (1957); Oberflache-Surface, **13**, 219 (1972).
13. R. H. J. Preston and S. Sanyal, *J. Appl. Chem.*, **6**, 26 (1956).

14. F. Mansfeld and J. V. Kenkel, *Corr. Sci.*, **16**, 111 (1976).
15. F. Mansfeld and J. V. Kenkel, *Corrosion*, **33**, 13 (1977).
16. F. Mansfeld, *Werkstoffe und Korrosion*, **30**, 38 (1979).
17. N. D, Tomashov, *Theory of corrosion and protection of metals*, The McMillan Co., New York, 1966, Chapter XIV.
18. P. J. Sereda, ASTM, Bulletin No. 246, 47 (1960).
19. P. J. Sereda, *Ind. and Eng. Chem.*, **52**(2), 157 (1965).
20. P. J. Sereda, ASTM STP 558, 7 (1974).
21. H. Guttman and P. J. Sereda, ASTM STP 435, 326 (1968).
22. H. Guttman, ibid. 213.
23. P. J. Sereda, S. G. Croll and H. F. Slade, ASTM STP on atmospheric corrosion (in press).
24. V. Kucera and E. Mattsson, ASTM STP 558, 239 (1974).
25. V. Kucera and J. Gullman, ASTM STP 727, 238 (1981).
26. F. Mansfeld and S. Tsai, *Corr. Sci.*, **20**, 853 (1980).
27. Yu. N. Mikhailovskii, G. B. Klark, L. A. Shuvakhina, A. P. Sanko, Yu. P. Gladkikh and V. V. Agafonov, *Prot. Metals*, **7**, 466 (1971).
28. G. Woelker, *Z. Meteorologie*, **13**, 30 (1972).
29. F. Mansfeld, Atmospheric corrosion monitoring. Final Report, Subcontract No. 31-109-38-4247, May 1981.
30. I. Jirovsky, D. Knotkova, I. Kokoska and J. Prosek, *Proc. Symp on Atmospheric Corrosion*, The Electrochemical Society, Hollywood, Florida, Oct. 1980 (in press).
31. C. Voigt, *Korrosion* (Dresden), **10**, 3, 179 (1979).
32. S. Haagenrud, *Werkstoffe und Korrosion*, **31** (1980).
33. F. Mansfeld, in *Advs. Corros. Sci. Tech.*, **6**, 163 (1976).
34. W. Friehe and W. Schwenk, *Stahl und Eisen*, **97**, 686 (1977).
35. F. Mansfeld, Evaluation of electrochemical techniques for modelling and monitoring of atmospheric corrosion phenomona. Final Report to National Science Foundation, Grant No. DMR-7923965, May 1981.
36. Yu. N. Mikhailovskii, L. A. Shuvakhina, and C. T. Hong Wang, *Prot. Metals*, **9**, 135 (1973).
37. Yu. N. Mikhailovskii, G. B. Klark, L. A. Shuvakhina, V. V. Agafonov and N. I. Zhuravleva, *Prot. Metals*, **9**, 240 (1973).
38. Yu. N. Mikhailovskii, P. V. Strekalov and V. V. Agafonov, *Prot. Metals*, **16**, 308 (1980).
39. F. Mansfeld and S. Tsai, unpublished results.
40. F. Mansfeld, *Corrosion*, **37**, 301 (1981).
41. F. Mansfeld, M. W. Kendig and S. Tsai, Application of interrupter technique for corrosion studies in low conductivity media, to be submitted to *Corrosion*.
42. K. Bohnenkamp, G. Bergmann and W. Schwenk, *Stahl und Eisen*, **93** (122), 1054 (1973).
43. F. Mansfeld, Regional air pollution study, effects of airborne sulfur pollutants on materials, EPA-600/4-80-007, U.S. Environmental Protection Agency, Research Triangle Park, N.C. 27711, January 1980.
44. F. Mansfeld, Results of thirty months exposure study in St. Louis, Mo. CORROSION/78 paper No. 88; *Proc. 7th Int. Cong. Metallic Corrosion*, Rio de Janeiro, Brazil, October 1978, paper No. 193.

Chapter 2

ATMOSPHERIC CORROSION OF NON-FERROUS METALS

The late V. E. CARTER
19 *Newsholme Lane, Waterfield,*
Yorkshire, UK

1. INTRODUCTION

Non-ferrous metals fall into two broad groups with respect to their corrosion behaviour in natural atmospheres.

Group 1. Those metals which tend to corrode freely and at substantially constant rates—e.g. zinc and cadmium.

Group 2. Those metals which have an inherently high resistance to corrosion as a result of naturally formed oxide films and whose initial, limited, rates of corrosion are substantially reduced as the period of exposure increases and the quality of the protective films is improved. Non-ferrous metals in this group include aluminium, copper, lead and nickel.

In all cases, however, the corrosion performance of the non-ferrous metals is greatly dependent upon the type of atmospheric environment to which they are exposed and to the climatic conditions prevailing during exposure. It is important, therefore, to classify the type of environment in which any given metal is to be used and/or tested and to obtain as full details as possible of climatic factors such as temperature (mean and variations), relative humidity (mean and variations), rainfall (quantity and frequency) and others.

1.1 Types of Environments

Many systems of classification of types of atmospheres have been evolved by workers in the field of corrosion over the years. All broadly divide up the environments into industrial, urban, rural and marine types generally, dependent upon the nature of the pollutants present and their concentration and hence their severity of action. Attempts are being made in the International Standards Organization to produce standard classifications of natural atmospheric environments but these have not currently progressed beyond the stage of initial draft proposals for discussion. There is, however, a classification of environments in British Standard BS 5493:1977[1] which defines four basic types of environment.

Type 1.   *Non-polluted inland.* This environment covers most rural and suburban areas characterized by low levels of atmospheric pollutants of all kinds.

Type 2.   *Polluted inland.* This environment differs from Type 1 in that the level of atmospheric pollutants is markedly increased and hence is usually applied to industrialized areas.

Type 3.   *Non-polluted coastal.* This environment is similar to Type 1 in that pollutants of industrial origin are absent but marine pollutants are present in this type of atmosphere. These regions are typically less than 3 km from the sea coast, the distance depending upon how the strength and direction of the prevailing winds and the local topography produce build-up of salt contamination in the atmosphere.

Type 4.   *Polluted coastal.* This environment consists of areas of similar location to those of Type 3 but with substantial amounts of atmospheric pollution derived from industrial sources in addition to the marine pollutants.

In addition to these four broad categories of environmental conditions it is often necessary to consider additional special classifications such as the following:

A.   *High rainfall areas.* In these areas exposed metals will be more frequently and more thoroughly wetted with consequent increases in corrosion rates.

B.   *Arid areas.* Rainfall and humidity are low in these areas and hence corrosion rates are markedly reduced.

C.   *Tropical areas.* In these regions excessive temperature effects may influence the corrosion rate and in addition rainfall in these areas may vary excessively either on a daily or a seasonal basis.

D. Abrasive areas. Wind-blown abrasive particles of sand or other material which can have an erosive effect are prevalent in these areas. Erosion may affect protective oxide films and/or any insoluble corrosion products which would otherwise build up on the metal surface.

Attention is drawn in BS 5493[1] to the need to consider other special conditions additional to the broad categorizing of the type of environment. These special considerations are as follows:

1. Are environmental changes likely to occur during the period of service or testing of the metal (e.g. development and consequent industrialization in the area with consequent increase in pollution)?
2. Are there any local sources of pollution which may increase the corrosivity beyond that expected for the general type of area in question (e.g. a source of sulphur pollution of the atmosphere which is situated to windward of the test or service location and which may affect localized pollution levels even though the source of pollution may be a substantial distance away from the site)?
3. Are sheltering effects likely to be present as a result of buildings or the local topography in the immediate vicinity of the site?

1.2 Climatic Factors

Probably the most important climatic factors affecting corrosion rates are the number and frequency of wetting of the exposed metals, the amounts of sulphur and chloride pollution in the atmosphere and the temperature ranges encountered. All these factors must be considered when judging the suitability of a given metal for service in a given location or when carrying out atmospheric exposure corrosion testing programmes. A standard (ISO/DIS 4542),[2] giving guidance on outdoor corrosion tests for coated metals, is currently in the course of publication and a very similar standard is in course of preparation for testing metals which have not been coated. These documents include the following list of climatic factors which characterize exposure conditions:

1. Air temperature
2. Relative humidity
3. Absolute humidity } maximum, minimum and average values required.
4. Hours of sunshine.

5. Precipitation—type, amount, duration pH, sulphur and chloride pollution.
6. Wind speed and direction.
7. Air contamination—sulphur and chloride.
8. Insoluble dust contamination—quantity and composition.

Data of this nature can be readily obtained from public meteorological records but it must always be borne in mind that the results are only totally applicable to the immediate vicinity of the monitoring station concerned and that local conditions may modify some of the factors (notably the pollution levels). For this reason it is always desirable to record as many as possible of the relevant climatic factors at an exposure testing site or to obtain data specific to any important service location.

2. OUTDOOR CORROSION TESTING

The gathering of corrosion data by atmospheric exposure is, of necessity, a very long-term procedure, particularly when the metals concerned are of the more corrosion-resistant type. Test programmes often need to be planned to cover periods up to 10 or more years and it is important, therefore, to ensure that provision is made at the outset to test the effects of as many variables as possible—e.g. different types of atmospheres, varied orientation of specimens and the effect of different angles of exposure. All these points are considered in ISO/DIS 4542.[2]

Considerable variation in corrosion resistance is often found between metal surfaces exposed facing skywards and those facing groundwards (or otherwise partially sheltered) since the rates of drying and the periods of wetness will differ for the two conditions. Separate data should be gathered to cover these aspects of exposure. This matter is of particular importance where localized (pitting) corrosion occurs or where sacrificial protective coatings are employed since the partially sheltered areas are frequently those which are most rapidly attacked and premature failure in service in such locations may result from lack of the necessary data being available to enable a proper choice of material to be made.

Initial weather conditions when exposure to the atmosphere commences can markedly affect the subsequent progress of corrosion. For example, aluminium alloys exposed initially in winter developed corrosion pitting which was 1·5–2 times as great as when exposed initially in the summer (as measured after each group had had one year's exposure) though the total weight losses for both groups of specimens were similar.[3] The difference is explained by those specimens first exposed in winter remaining covered with a film of moisture for a greater proportion of the time during the first few weeks or months on test thus allowing the corrosion pits to remain active for longer periods. Separate groups of specimens exposed at different seasons are often desirable, therefore, and ISO/DIS 4542[2] also mentions this point, recommending that outdoor testing should, when possible, commence either in the April—May or the September—October period. These initial variations in corrosion rates, however, are of lesser importance as the total exposure period is extended due to the subsequent development of protective oxide films or insoluble corrosion products, provided always that the initial effect is not so great as to lead to premature failure.

The degree and frequency of maintenance during exposure can also be of importance. For example, the performance of anodized aluminium can be markedly improved by regular washing down to remove corrosive deposits—particularly in heavily polluted atmospheres where such deposits remain acidic and retain the corrodents in contact with the anodized surface.[4-6]

Another factor which may be of importance in affecting corrosion performance is that of relative movement of the materials exposed. Thus, when metals protected by electroplated coatings of nickel plus chromium, in which the chromium topcoat is of the microdiscontinuous type, are exposed to polluted atmospheres under conditions of static exposure excessive dulling of the bright chromium surface occurs. However, when such finishes are used in service on motor vehicles (their principal service usage) this heavy surface dulling is suppressed for long periods.[7] The reason for the difference in behaviour is probably associated with the different corrosive conditions encountered as a result of the movement of the vehicle resulting in different periods of wetness from those encountered in static exposure and the presence of road washes which produce different corrodent electrolytes on the metal surface. It is sometimes necessary, therefore, to supplement static exposure with mobile testing to obtain data relevant to particular service conditions.[2]

3. ACCELERATED CORROSION TESTING

Because of the long periods of time which are required in order to gather data on atmospheric exposure, attempts may be made to obtain information more rapidly by the use of accelerated corrosion tests in the laboratory. Most of the standard accelerated corrosion tests involve exposure to increased humidity and often raised temperature with or without the addition of a corrodent fog or a gaseous sulphur dioxide pollutant[8-12] or alternatively by the application of a corrodent paste.[13] However, because these accelerated tests provide entirely different conditions of wetness and dryness from those encountered in outdoor exposure, and also different quantities of corrodent are available, the effect on the production of corrosion products is such that the nature of the corrosion process may be altered. Consequently, results obtained in accelerated tests are, at best, only related directly to outdoor usage by a correlation factor which may be established sometimes by comparing the results of several series of accelerated tests with those obtained in outdoor exposure tests on replicate specimens.[14,15] At worst, the results of accelerated tests may only be used for sorting on a 'go, no go' basis for acceptance purposes. It cannot be too strongly emphasized that the prediction of service life from data obtained in accelerated corrosion tests is extremely unreliable. Furthermore, accelerated tests are not suitable for use with metal-coated systems in which the coating provides sacrificial protection to the substrate metal; warnings to this effect are included in ASTM Standards A164–55[16] and A165–55[17] for electrodeposited coatings of zinc and cadmium on steel.

4. CORROSION OF INDIVIDUAL METALS AND ALLOYS

4.1 Aluminium and Aluminium Alloys

In its pure form aluminium offers a high degree of corrosion resistance to the atmosphere because of the rapid formation on exposure to air of a thin, tenacious oxide film. This oxide film is inert and when it is present further corrosion of the metal surface is stifled. Such corrosion as does occur takes place at defective points in the oxide film and takes the form of localized pitting instead of general surface wastage.

In an industrial environment the corrosion rate of aluminium, averaged over a six year period, is between 2 and 5 μm per year[18] but the corrosion rate in the sixth year is only one-quarter that which occurs in the first year. Although the corrosion rate averaged over the whole surface is so low the rate of local penetration is much higher because the corrosion is confined to a number of small pits initiated at defects in the oxide film. Typically these pits may reach a depth of 0.25–0.5 mm after about six years' exposure to an industrial environment.[18] However, pitting to this depth has little, if any, effect on the overall strength of the material except where thin section sheet is concerned. Even with thin sheet the loss of strength is limited but the danger of failure due to penetration may be substantial.

When exposure of pure aluminium is to a marine or rural environment the average corrosion rate is reduced to about one-tenth of the rate which occurs when exposure is to an industrial environment and the depth of penetration of pitting is reduced to about one-fifth.[18]

Although pure aluminium has a very low corrosion rate the metal is soft and of low mechanical strength which limits its applications. The strength can be greatly increased by alloying but the addition of some alloying elements severely reduces the corrosion resistance of the alloy when compared with that of pure aluminium.

Aluminium–magnesium alloys, often with the addition of silicon and/or manganese as additional alloying elements mostly offer moderate to good corrosion resistance coupled with moderate strengths (150–300 N/mm^2). High strength (300–350 N/mm^2) can be achieved with the aluminium–copper–magnesium alloys and the aluminium–zinc–magnesium alloys but with these materials the corrosion resistance is poorer than that of many of the lower strength alloys.

The poorer corrosion resistance in high-strength alloys is particularly true of those aluminium alloys containing substantial alloying additions of copper. These alloys may suffer a special form of corrosion known as 'exfoliation corrosion' (sometimes known as 'layer corrosion'). This form of corrosion has been described by Sutton[19] who states that when it occurs the corrosion spreads along planes between lamellae lying parallel to or in the direction of deformation during rolling or extrusion. Bell and Campbell[20] add that the corrosion products produced as a result of this selective intergranular attack lift layers of uncorroded metal from the surface and subsequent corrosion of these exfoliated layers produces the appearance which is typical of this form of attack (see Fig. 1). Susceptibility to exfoliation corrosion is very dependent upon the nature of the microstructure of the metal, its initiation being delayed or prevented by the presence of an equiaxial recrystallized layer adjacent to the exposed surface. Modifications to the metal working procedures can, therefore, markedly influence susceptibility to exfoliation corrosion— notably, over-ageing can markedly reduce the incidence of attack[20] while fabrication processes such as welding can result in highly susceptible regions particularly in the heat-affected zone of welds.[19]

FIG. 1. Superficial pitting (lhs) and exfoliation corrosion (rhs) of aluminium–copper–magnesium alloy. (Magnification × $\frac{1}{2}$.) Reproduced by kind permission of BNF Metals Technology Centre.

Exposure data for various aluminium alloys exposed at a variety of sites in different parts of the world are given in papers by McGeary *et al.*,[18] Lindgren *et al.*[21] and Kalinin[22] and have been abstracted and summarised in Table 1.

Atteras and Hagerup[23] exposed pure aluminium at a number of different sites in Norway and their results show that marine atmospheres in western Norway produce less corrosion of aluminium than is found to occur in exposure at marine sites in more southerly latitudes.

The effect of climatic factors in the corrosion of aluminium alloys can also be judged by comparing the penetration rates of an aluminium–copper–magnesium alloy exposed at marine and rural sites in temperate and tropical regions. A penetration rate of 2 μm per year in temperate marine exposure[18] contrasts with a penetration rate of 6 μm per year in a tropical marine environment[24] while a rate of 0.2 μm per year in a temperate rural environment[18] contrasts with 0.6 μm per year in a tropical rural environment.[24]

The effects of climatic factors on the corrosion of aluminium have been studied in some detail by Mikhailovskii[25] who found that the corrosion rate increases linearly with the length of time over which the metal is exposed to relative humidities in excess of 80% and also with the length of time during which liquid moisture films are present on the metal surface. After determining corrosion rates under adsorbed and condensed films of moisture when exposed to an unpolluted rural environment they produced a formula which enables calculation of the factors for the acceleration of corrosion due to sulphur and chloride pollution in the atmosphere and thus the calculation of corrosion rates for any climatic region for which the basic data are available. Comparison of some calculated and observed corrosion rates is shown in Table 2.

The particularly adverse effect of copper additions on the corrosion resistance of aluminium is due to the bimetallic effect between copper and aluminium which causes catastrophic accelerated attack on the aluminium if the two metals are in direct metallic contact. In the case of copper as an alloying element the loss of corrosion resistance appears to be due to precipitation of metallic copper particles from solution when copper-rich constituents of the alloy microstructure are attacked by a corrodent. The metallic copper particles are deposited on aluminium metal surfaces and set up the bimetallic cells which produce preferential attack on the aluminium. This is clearly revealed by the blue coloration of the corrosion products which are produced when copper-bearing aluminium alloys corrode. It is also sometimes possible to detect the presence of metallic copper either by microscopic examination or by the use of microchemical analysis techniques.

Although it is well known that aluminium and copper alloy components should never be in direct metallic contact with each other it is surprising how often such cases occur in practice with consequent adverse effects in service. It is perhaps less well known that enhanced corrosion of aluminium may be caused by the run-off of rain from adjacent copper components not in direct metallic contact with the aluminium. Slight corrosion of the copper component may allow sufficient copper salts to become dissolved in the run-off liquid to produce a precipitate of metallic copper on the aluminium on to which it drips. In experimental work by Sick[26] pure aluminium was exposed for six years to rainwater contaminated with copper derived from a copper panel suspended just above the aluminium panel. Comparison of the corrosion of the aluminium exposed to the copper-containing water with the corrosion of a similar aluminium panel exposed at the same time to copper-free water showed that the degree of superficial corrosion of the aluminium was greater when exposed to the copper-containing water though the difference was not sufficient to affect its functional perfor-

TABLE 1
ATMOSPHERIC CORROSION DATA FOR ALUMINIUM ALLOYS

Alloy type	Condition	Exposure site Type	Country	Period of exposure (years)	Max. depth of attack (μm)	Type of attack	Lit. ref. no.
Al–2.5Mg	Welded Sheet	Marine	Sweden	10	90	Pitting & slight exfoliation in HAZ.	21
Al–2.5Mg	Sheet	Marine	USA	7	90	Pitting	18
Al–2.5Mg	Sheet	Ind.	USA	7	90	Pitting	18
Al–3.5Mg	Welded Sheet	Marine	Sweden	10	120	Pitting & slight exfoliation in HAZ.	21
Al–3.5Mg	Sheet	Marine	USA	7	130	Pitting	18
Al–3.5Mg	Sheet	Ind.	USA	7	100	Pitting	18
Al–5Mg	Welded Sheet	Marine	Sweden	10	90	Pitting & slight exfoliation in HAZ.	21
Al–5Mg	Welded Extrusion	Marine	Sweden	10	160	Pitting & slight exfoliation in HAZ	21
Al–5Mg	Sheet	Marine	USA	7	110	Pitting	18
Al–5Mg	Sheet	Ind.	USA	7	130	Pitting	18
Al–5Mg	Sheet	Ind.	UK	6	575	Intergranular pitting†	18
Al–5Mg	Sheet	Marine	UK	6	110	Intergranular pitting†	18
Al–5Mg	Sheet	Rural	UK	6	25	Intergranular pitting†	18
Al–Si–Mg	Welded Sheet or Extrusion	Marine	Sweden	10	100	Pitting & slight exfoliation in HAZ	21

Alloy	Form	Environment	Country			Corrosion type	Ref
Al–Si–Mg	Sheet	Marine	UK	6	90	Intergranular pitting†	18
Al–Si–Mg	Sheet	Ind.	UK	6	475	Intergranular pitting†	18
Al–Si–Mg	Sheet	Rural	UK	6	40	Intergranular pitting†	18
Al–Cu–Si	Sheet	Ind.	USSR	5	100	Intergranular pitting†	22
Al–Cu–Si	Sheet	Marine	USSR	5	100	Intergranular pitting†	22
Al–Cu–Si	Sheet	Rural	USSR	5	50	Intergranular pitting†	22
Al–Cu–Mg	Sheet	Ind.	USSR	5	210	Exfoliation	22
Al–Cu–Mg	Sheet	Marine	USSR	5	80	Exfoliation	22
Al–Cu–Mg	Sheet	Rural	USSR	5	100	Exfoliation	22
Al–3.5Zn–Mg	Sheet	Ind.	USSR	5	150	Intergranular pitting†	22
Al–3.5Zn–Mg	Sheet	Marine	USSR	5	50	Intergranular pitting†	22
Al–3.5Zn–Mg	Sheet	Rural	USSR	5	80	Intergranular pitting†	22
Al–5Zn–Mg	Welded Sheet or Extrusion	Marine	Sweden	10	70	Pitting & exfoliation in HAZ	21
Al–5Zn–Mg	Sheet	Marine	USA	7	110	Pitting	18
Al–5Zn–Mg	Sheet	Ind.	USA	7	120	Pitting	18
Al–5Zn–Mg	Sheet	Rural	USA	7	45	Pitting	18
Al–Zn–Mg–Cu	Sheet	Ind.	USSR	5	100	Intergranular pitting†	22
Al–Zn–Mg–Cu	Sheet	Marine	USSR	5	100	Intergranular pitting†	22
Al–Zn–Mg–Cu	Sheet	Rural	USSR	5	50	Intergranular pitting†	22

Summarized and abstracted from Refs. 18, 21 and 22.

† Corrosion initiated preferentially at grain boundaries followed by spreading of the attack into the grains which may subsequently become converted to corrosion products.

TABLE 2

COMPARISON OF CALCULATED AND OBSERVED
CORROSION RATES FOR ALUMINIUM

Atmosphere	Corrosion rate (g/m^2)	
	Calculated	Observed
Tropical-coastal	0.30	0.15–0.40
Rural sub-tropical	0.12	0.21–0.40
Coastal sub-tropical	0.48	0.4–0.7
Rural non-tropical	0.20	0.16–0.18
Industrial non-tropical	0.48	0.45
Coastal frigid	8.3	5.3–9.2

Summarized and abstracted from Ref. 25.

mance significantly. Examination of the corroded aluminium showed that redeposited copper particles were embedded in aluminium corrosion products which rendered them inactive to the aluminium matrix and chemical analysis revealed that the copper concentration present was 0.025 mg/cm^2 of aluminium surface exposed.

Enhanced corrosion of aluminium may also be caused through bimetallic contact with bare steel. This is generally less dangerous than contact with copper since the potential difference between steel and aluminium is only small but it is nevertheless advisable to avoid direct metallic contact between steel and aluminium. Alternatively, when the two metals must be coupled together protective measures should be employed to prevent enhanced attack on the aluminium, for example by applying a cadmium coating to the steel and sealing the joint to prevent the ingress of moisture.

Contact between aluminium and carbon can also lead to accelerated attack on the metal and cases of premature failure of aluminium roofing sheet have occurred because of carbon particles deposited as soot in the lee of chimneys of inadequate height.

A feature of aluminium is its high degree of susceptibility to crevice corrosion in joints, fasteners or other overlapping applications. This is the result of differential aeration cells being set up in the moisture films which are retained in the joint. The formation of the natural protective oxide film is hindered in the low-oxygen regions of the cell and enhanced corrosion of the aluminium metal occurs in these regions. Here again protection is required by the exclusion of corrodent moisture from all joints by the use of suitable inert mastics.

Aluminium–zinc–magnesium alloys of the highest strength may be susceptible to stress corrosion leading to premature failure in highly stressed applications. As with susceptibility to exfoliation corrosion some degree of control may be exercised by particular attention to alloy composition and microstructure;[27] the problem may also be aggravated by the use of welding processes.[28]

4.1.1. Protection of Aluminium

Some protection against corrosion of aluminium and its alloys may be achieved by means of chromate conversion coatings as a result of the inhibitive action of the chromate complexes formed on the metal surface together with the limited degree of physical exclusion of the corrodent liquid obtained by the deposited chromate film. Pearlstein and Teitell[24] found that 98% protection of aluminium was obtained by chromating when exposure was for one year in a tropical marine environment and 92% protection was obtained when the chromated aluminium was exposed for one year in a tropical open field environment. However, the chromate film is very thin and has a poor resistance to abrasion so that protection is generally only retained for a limited period. The principal use of chromate treatments is as pre-treatment procedures prior to painting, the chromate film providing an improved key for better ad-hesion of the paint film while retaining inhibitive chromates at the paint/metal interface which may delay breakdown when moisture per-meates the paint coating.

Apart from protection by chromates and paint coatings a high degree of protection against corrosion of aluminium alloys due to their alloy composition—such as exfoliation or stress corrosion—can be achieved by applying metallic coatings. Alloys may be clad with pure aluminium by rolling or extrusion processes or alternatively coatings of aluminium or zinc may be applied by metal spraying processes. The principle on which the clad or sprayed metal coatings protect the alloy core metal is the dual one of the intrinsically higher corrosion resistance of the metal coating material and the sacrifical protective action achieved on the more highly alloyed core metal when it is exposed by corrosion penetrat-ing the coating. Thus, McGeary et al.[18] achieved a 20–40% reduction in corrosion rate for clad aluminium alloys over their unclad counterparts in seven years' atmospheric exposure in the USA. Aluminium–copper–magnesium alloys protected by 125 μm thick sprayed coatings of pure aluminium were found by Carter and Campbell[29] to retain full mechanical strength over 10 years' atmospheric exposure in the UK while unprotected alloys lost 33% strength through exfoliation corrosion. Similarly, sprayed coatings of aluminium or of zinc 125 μm thick completely prevented stress corrosion of aluminium–zinc–magnesium alloys over a 10 year period whereas the unprotected alloys failed by stress corrosion in less than 2 years.[30]

Probably the principal means of protecting aluminium against corrosion is by means of the application of an anodic oxide film by sulphuric acid anodizing. The electrolytic coating process provides a transparent oxide film which allows the decorative effect of the aluminium metal surface—either as-wrought, brightened or etched—to be retained unimpaired for long periods of exposure to atmospheric environments. Alternatively, colour may be imparted to the coating by means of organic dyestuffs absorbed in the anodic oxide film, by electrolytic deposition of metallic particles in the film, or by developing integrally coloured films by using special anodizing processes.

The efficacy of protection achieved by anodizing is dependent upon the thickness of the anodic oxide film and the quality of the sealing process applied to the porous film which is produced by the anodizing process. The severity of the environment to which the anodized material is exposed, the degree of maintenance applied during service and the chemical composition and metallurgical condition of the anodized metal also affect their performance in outdoor service. By applying well-sealed anodic oxide coatings of 25 μm minimum thickness to suitable alloys service lives of 10–20 years in severe industrial environments with negligible deterioration from the perfect initial appearance can be readily achieved.[31,32]

Anodizing may also be used as a pre-treatment for painting but for this purpose the anodic oxide film—produced either by the sulphuric acid process or by the chromic acid process—which is applied is limited in thickness to 1–2 μm and is left unsealed so as to provide the best adhesion of the paint film.[5]

An unusual example of failure of anodized aluminium in service during exposure to an industrial atmosphere has been investigated by Wettinck.[33] Failure by the development of a heavy white staining of the transparent anodic oxide film without the development of pitting corrosion was shown to be caused by fluoride contamination of the atmosphere to which it had been exposed. However, this is a very special case which is not likely to be encountered in normal service.

4.1.2 Protective Coatings of Aluminium

Apart from their use as protective coatings for aluminium alloys (see above) aluminium coatings may also be used for the protection of steel. Coatings are generally applied either by metal spraying or by hot-dipping and are highly resistant to atmospheric corrosion. A 100 μm thick sprayed aluminium coating has successfully prevented rusting of steel exposed to an industrial environment in the UK for a period of seven years compared with rusting in five years with a 75 μm thick sprayed zinc coating and as little as one and a half to two years with a 25 μm thick hot-dip galvanized zinc coating. Protection is due to the build-up of thick, dense aluminium oxides in the pores of the coating which stifle further corrosion and possibly also to some limited sacrificial protection of the steel by the aluminium coating.

4.2 Magnesium Alloys

Magnesium alloys exposed to the atmosphere remain free from corrosion during periods when the relative humidity remains below about 60%[34] but scattered superficial corrosion spots develop when the humidity approaches 100%. Little difference in behaviour is seen between the various magnesium-base alloys and in two years' exposure to an industrial environment in the UK the maximum rate of penetration measured was 76 μm per year.[34] The air-formed corrosion products in industrial environments consist of a mixture of carbonates, sulphates and hydroxides but in marine atmospheres chlorides and oxychlorides are present which, being hygroscopic, tend to produce increased corrosion rates.

Rybakov et al[35] found that the localized centres of corrosion on cast magnesium alloys were related to the local accumulations of casting and refining fluxes since chlorides of potassium and calcium were detected at the corrosion sites.

Brandt and Adam[36] exposed a range of magnesium-base alloys in both the sand-cast and wrought condition at atmospheric sites in the USA. They found that the corrosion, measured by loss of tensile strength, did not differ greatly with the alloy composition, was generally linear in rate or somewhat increasing with time of exposure, and tended to be more severe in the heat-treated condition than in the as-cast condition. In a 10 year exposure test series the range of losses of tensile strength were as given in Table 3. The authors state that, in general, the magnesium-base alloys suffer greater loss of strength than do aluminium alloys.

Pearlstein and Teitell[24] also report linear corrosion/time curves for exposure of magnesium alloys at tropical marine, open field and rain forest sites during a four year period, basing their results on measurements of weight loss.

TABLE 3

LOSS OF TENSILE STRENGTH FOR MAGNESIUM ALLOYS EXPOSED FOR 10 YEARS IN THE USA

Condition	Exposure site	Percentage loss of tensile strength
Cast alloys	Industrial	6–27
	Marine	6–27
	Rural	4–22
Wrought alloys	Industrial	12–20
	Marine	3–12
	Rural	4–18

Summarized and abstracted from Ref. 36.

The bimetallic effect of coupling magnesium alloys with steel results in some increase in the rate of corrosion of the magnesium, even if the steel has been plated with zinc or cadmium. The increase in corrosion, however, is not excessive when normal atmospheric exposure is involved, but is more marked in conditions of high humidity and high chloride contamination.[34]

4.2.1 Protection of Magnesium Alloys

The methods of protection for magnesium alloys are similar to those employed for the protection of aluminium and its alloys. Thus, chromating may be used to provide a limited degree of protection and as a pretreatment for painting. Pearlstein and Teitell[24] exposed bare and chromated magnesium alloy panels in tropical marine, open field and rain forest environments and during a one year exposure period the chromated panels achieved 50%, 13% and 22% protection respectively at these three test sites; the percentage protection achieved was markedly less than that achieved with chromating of aluminium panels exposed at these sites during the same test period.

Anodic oxidation processes may also be applied to magnesium alloys but, unlike the anodized films produced on aluminium alloys, those on magnesium alloys do not provide effective long-term protection to the metal and hence should be considered to fall more within the category of temporary protection methods and used instead as pre-treatments for paint finishes.[34] Furthermore, whereas anodized aluminium alloys have an attractive clear finish, anodized magnesium finishes have an uneven, drab brown or greenish coloration. The two anodizing processes mainly used for magnesium alloys are either a potassium hydroxide/sodium phosphate/potassium fluoride bath or an ammonium bifluoride/sodium dichromate/phosphoric acid bath.[34]

4.3 Titanium and Titanium Alloys

Titanium and its alloys are highly resistant to corrosion on exposure to the atmosphere due to the fact that a thin, tenacious surface film of titanium dioxide is rapidly formed on exposure to damp air. Greenlee and Plock[37] exposed eight titanium alloys representative of the alpha, alpha-beta and beta types for periods up to seven years at industrial, rural and marine sites in the USA during which time only slight staining and isolated local areas of discoloration occurred without general or pitting corrosion or any loss of mechanical properties.

Resistance to crevice corrosion is greater than with most other non-ferrous metals and titanium is generally not itself adversely affected by bimetallic contact with other metals. However, bimetallic couples involving titanium may result in enhanced attack on the second metal of the couple system since titanium acts as a very efficient cathode.

Owing to the high degree of corrosion resistance of titanium and its alloys methods of protection do not normally need to be applied. However, anodizing of titanium has been developed commercially, principally in order to obtain special decorative effects since the anodic oxide films can be produced in a wide range of different colours, dependent upon their thickness and the methods of control of the anodizing processing parameters as well as the metallurgical condition of the alloy treated.

Protective anodizing of titanium and its alloys using a sulphuric acid process may be applied as an anti-galling treatment for fasteners, as a pre-treatment for solid film lubricant applications or to provide for the protection of less-noble metals used in contact with the titanium.

4.4 Zinc and Zinc Alloys

Zinc corrodes freely but relatively slowly in the atmosphere at an essentially constant rate. The corrosion rate in an industrial atmosphere is about 15 μm per year and falls to about one-fifth of that value in a marine or rural environment.[38] The reason for the low corrosion rate is the tendency to produce basic zinc chloride and carbonate corrosion products which tend to stifle the attack. When industrial pollution by sulphur gases is present in the atmosphere the acidic condensates produced on the corroding metal surface increase the rate of dissolution until sufficient basic salts have been produced to provide further protection and so reduce the rate of attack again; under these conditions some basic sulphates may also be produced. Further deposition of acidic condensates produces a fall in pH and the rate of attack again increases.[34] The combined effect of wetness and atmospheric pollution on the corrosion rate of rolled zinc has been studied by Guttman,[39] as shown in Fig. 2, and the effect of wetness on the corrosion rate is also

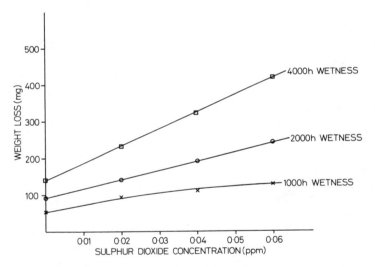

FIG. 2. Effect of sulphur pollution and total time of wetness on corrosion weight loss of zinc in atmospheric exposure. (Abstracted from Ref. 39.)

reflected in his numerical results which consistently showed greater corrosion rates for skyward facing surfaces than for groundward facing surfaces.

Schickorr[40] exposed zinc to the atmosphere in Stuttgart and determined analytically the amount of zinc present each month in the corrosion products on the surface of the metal as well as in the collected run-off of rainwater from the metal surface. These figures were compared with the total weight losses of the metal panels and the results show variations in corrosion rate month by month with considerably higher values in winter than in summer.Furthermore, the quantity of corrosion products adhering to the surface of the metal was found to be related to monthly variations in rainfall and the total amount of zinc determined in the corrosion products agreed closely with the amount of the weight loss of the zinc panels.

Studies of the effect of atmospheric pollution by chloride on the corrosion rate of zinc[41] show that in tropical exposures in Nigeria there was an approximately linear relationship between the corrosion rate and salinity but there was no discernible correlation of these factors in exposures carried out in the UK.

Unlike aluminium (as noted above) the month in which corrosion tests commence has little effect on the yearly corrosion rate according to further work by Schickorr[42] with tests at four different sites, whereas Sanyal et al.[43] did find some effects due to the time at which exposure was started at some sites in India.

Gilbert[44] found that when a 2% copper alloying addition was made to pure zinc the corrosion rate was reduced by about 20% when exposure was made to industrial environments in the UK but not when the exposure was to a UK marine environment. Dunbar,[45] however, found that at both industrial and marine sites in the USA the weight losses of the two materials were similar although localized pitting occurred to a greater depth on the copper-bearing material. Adverse effects of copper alloying additions were also shown in 20 year exposures of zinc diecasting alloys at industrial and rural environments in the USA[46] where the copper-bearing alloys lost 31% and 15% of tensile strength respectively and the comparative figures for the copper-free alloys were 19% and 12%.

4.4.1 Protection of Zinc

Under conditions of high humidity zinc suffers a special form of superficial corrosion known as 'white rusting'. This is characteristic of exposure under conditions which provide long periods of wetness with relatively uncontaminated moisture in which the white corrosion products of zinc produced by the corrosion adhere to the metal surface to give the typical white stained appearance. Limited protection against white rusting and also against general initial atmospheric corrosion can be achieved by the application of chromate passivation coatings as with aluminium alloys. Chromate passivation treatments may also be used as pre-treatments for painting.

Where it is desired to retain a particularly high degree of surface finish on zinc for long periods lacquer films may be applied to exclude atmospheric corrodents. Many lacquers, however, suffer rapid oxidative breakdown by the action of the ultra-violet component of sunlight which causes yellowing and opacity. The acidic breakdown products can also readily attack the metal surface. By careful choice of the type of lacquer and the incorporation in the lacquer formulation of ultra-violet absorbers which delay oxidative breakdown, plus inhibitors of the chelating type which prevent attack on the metal, suitable lacquers have been produced which retain clarity and anti-tarnish protection for extended periods. Christie and Carter[47] have described the performance of lacquers of this type using additives of the benzotriazole and benzophenone groups and with rubeanic acid added as the chelating agent for zinc. Using this type of formulation with both acrylic and polyurethane clear lacquer bases, protection of zinc for periods in excess of three years' exposure to industrial and marine environments in the UK has been obtained.

The protection of zinc alloy diecastings used in decorative applications in service is normally achieved by the use of protective and decorative nickel plus chromium coating systems described later in this chapter.

4.4.2 Protective Coatings of Zinc

Probably the greatest use of zinc in outdoor service is as a protective coating metal for other metals to which it can offer sacrificial protection. Mention has already been made earlier in this chapter to its use as a sprayed metal coating for protecting the strong aluminium alloys against exfoliation and stress corrosion. Electrodeposited and sprayed metal coatings of zinc are widely used for protecting steel against corrosion—either alone or in combination with suitable paints; some protective coatings of zinc are also applied by sherardizing. By far the largest amount of zinc coatings used on steel are applied by the hot-dip galvanizing process.

The protective value of zinc coatings on steel varies little with the way in which it is produced, corrosion rates being similar for both hot-dipped and electodeposited zinc. With sprayed zinc coatings there is some evidence to suggest that the protective value is enhanced by the nature of the coating which allows corrosion products to be readily retained in the coating pores so that further attack is stifled. With electrodeposited zinc Hippensteel and Borgmann[48] and Biestek[49] have reported that deposits obtained from a sulphate electrolyte perform better than those from a cyanide electrolyte though the differences are small.

With hot-dipped galvanized coatings of zinc on steel the level of performance is affected to some extent by the nature and thickness of the alloy layer formed between the outer (pure zinc) layer of the coating and the steel substrate. This is dependent upon the conditions of galvanizing and on any metallic alloying additions to the molten zinc galvanizing bath. Results have been summarized in a publication by Radeker et al.[50] which showed that addition of copper to the bath improved weathering resistance, aluminium decreased weathering resistance and cadmium or tin had little effect.

Work done by Hadden[51] showed that post-galvanizing annealing treatments substantially improved the outdoor corrosion performance of galvanized coatings as a result of the diffusion of iron into the zinc coating during the annealing process. This work was extended by Campbell et al.[52] to include tests on sprayed, electrodeposited and hot-dipped zinc coatings in which the corrosion performance was improved by heat treatments chosen so as to produce iron contents in excess of 15% in the coatings.

The general length of life obtainable in service in atmospheric environments, judged on the criterion of freedom from heavy rusting of the zinc-coated steel, may range from 5 to 50 years according to the thickness of the zinc coating and the severity of the atmospheric environment to which it is exposed.

Electrodeposited zinc coatings are generally confined to the lower thickness range (up to 20–30 μm) owing to economic considerations. Sprayed zinc coatings may be typically up to about 100 μm in thickness and galvanized coatings may be applied up to about 200 μm in thickness if special processes and grades of steel are used for the maximum thickness levels.

By combining metallic zinc coatings with superimposed paint schemes the protection of the steel can be extended to virtually unlimited periods provided that the paint schemes are well chosen and properly maintained. Data on these combined protective schemes are given in BS 5493.[1]

4.5 Cadmium

Cadmium is a metal with similar corrosion performance to that of zinc but the metal and its vapour are very toxic. Its use is confined to that of a protective coating metal applied to steel by electrodeposition. Like zinc its protective value is dependent upon its thickness but coating thicknesses which are applied do not normally exceed 20–30 μm because of the high cost of the metal.

Cadmium shows a marked superiority over zinc as a protective coating for steel when tested in a salt fog environment but this superiority is often not achieved in exposure to marine environments. The choice between cadmium and zinc as a protective coating metal for steel is a difficult one to make and is very markedly dependent upon local environmental considerations. A review of the subject and its literature made by Carter[53] shows that zinc is to be preferred in atmospheres where sulphur pollution is present but in marine environments the choice is dependent upon variations in humidity and the level of chloride contamination. In very high humidity locations, especially when condensation occurs freely, cadmium is the preferred choice. Cadmium is also free from white rusting. Apart from the strict technical merits governing a choice between cadmium and zinc, account must often be taken of two other considerations. Because of the high toxicity of cadmium its use may have to be ruled out even though its performance would be superior to zinc. Also, because of the high cost of cadmium it may well be economically more favourable to use a thicker protective coating of zinc to obtain equivalent performance to that of a thinner coating of cadmium in an application in which the performance of the latter metal is superior, thickness for thickness.

Variations in relative performance of cadmium and zinc may be explained by the nature of the corrosion products formed and the effects of condensation. Layton[54] suggests that soluble sulphates of both zinc and cadmium which are formed in industrial atmospheres will be freely removed by rain so allowing corrosion to continue unhindered. In rural and marine environments, however, carbonates and chlorides are formed; with cadmium these salts are less soluble than with zinc and so they are less easily removed by rain and can exert a greater stifling action on the continuing corrosion reaction.

Cadmium is the preferred coating metal for steel in applications where bimetallic contact with aluminium is involved and should always be specified in applications where these two metals must be joined together during assembly.

As with zinc, cadmium may also be protected to a limited degree and for a limited period by the application of chromate passivation treatments and these treatments are also used as pre-treatments prior to painting.

4.6 Copper and Copper Alloys

Copper and copper alloys are highly resistant to atmospheric corrosion due to the formation in air of a dark surface film consisting principally of cuprous oxide together with basic salts. The corrosion attack is uniform over the surface of the metal and the penetration rate ranges from 0.2 to 0.6 μm per year in rural environments to 0.9 to 2.2 μm per year in industrial environments.[55] Corrosion rates for copper alloys may be lower than that of pure copper as a result of the modification of film formation by the particular alloying element present; for example, the addition of aluminium to a brass or bronze alloy results in the production of a more protective oxide film in which the concentration of aluminium present is greater than the concentration of aluminium in the bulk alloy metal.

After several years' exposure to marine or industrial environments the principal salts present in the corrosion products produced on copper are basic copper sulphates with perhaps smaller amounts of basic copper carbonates. Some basic copper chloride may also be present when exposure has been to environments in which chloride contamination is present. The basic sulphates are derived from the action of sulphur pollutants in the environment and their presence leads to the development of the familiar green patina seen on architectural copper items such as roofs and gutterings. This green patina is decoratively acceptable and when it has become fully established proves to be very inert so that very long service lives are then achieved. Because of the long period of time needed to fully establish this patina, as well as the fact that the appearance of the article is streaky and less decoratively acceptable during its formative period, many attempts have been made over many years to accelerate the formation of a complete patina in service or to produce it artificially before putting the article into use.[56,57] Many of the studies have concentrated on the identification of the various components of the corrosion products formed in humid atmospheres containing sulphur dioxide.[58-60] Perhaps the best of the artificially produced patinas is a factory-applied process developed in Sweden and described by Mattsson and Holm.[61]

Vernon[62] found that the initial weather conditions on exposure influenced the degree of protection afforded to copper by the air-formed film. Under winter conditions when high levels of sulphur pollution were present, non-protective films were produced resulting in a linear corrosion rate while in summer, when sulphur pollution levels were low, protective films were rapidly formed which resisted any subsequent increase in pollution levels.

If pollution by hydrogen sulphide or organic sulphides is present in the atmosphere rapid tarnishing of copper will develop, even in the absence of high moisture levels, the tarnish being of a dark blackish brown coloration.

Mattsson and Holm[55] report results of two year and seven year exposures of 22 different coppers and copper alloys in urban, rural and marine environments in Sweden. The rate of attack, measured by weight loss and reported as average penetration rates per year, varied surprisingly little over the whole range of the materials tested. Comparison of the two year and seven year results showed that the corrosion rate was diminishing with time, the decrease being lowest in the urban and rural environments and greatest in the marine environment. Beta brass was the alloy which had the greatest corrosion rate in these tests and in the urban atmosphere the rate of penetration of this type of alloy increased with the zinc content. Losses in tensile strength were negligible for all the alloys tested except again for the brasses in which loss of tensile strength ranged from 12% to 32% after seven years' exposure.

Thompson[63] has also reported results of two year and seven year atmospheric exposure tests on a range of copper alloys carried out at industrial, marine and rural sites in the USA. In general his results were in line with those obtained by Mattsson and Holm[55] in Sweden and showed the greatest severity of attack at the industrial site with the lowest severity occurring at the rural site. The data he obtained were analysed statistically and the alloys arranged in an order of merit of corrosion resistance for each type of environment as shown in Fig. 3. It is interesting to note from these results that the alloy containing 2% silicon and 7% aluminium is consistently the best in all three types of environment and the alloy containing 0.5% beryllium and 2.5% cobalt is nearly as consistently the worst at all three sites.

A particular hazard with brasses is a special form of selective corrosion known as dezincification in which the alloy is attacked with redeposition *in situ* of the copper, the latter being redeposited in the form of a spongy mass which is porous and of very low mechanical strength. In single phase alpha alloys the attack can be fully inhibited by the addition of about 0.05% arsenic to the alloy composition, but the addition of arsenic is ineffective in preventing dezincification attack on duplex-type alloys. In the duplex alloys dezincification occurs first on the second phase solid solution but, in severe cases, it may spread later into the alpha phase. The attack is most likely to occur under immersed conditions in seawater or in natural waters of particular compositions but it may also develop in atmospheric exposure, often in areas of low oxygenation such as joint crevices or under the heads and on the threads of fasteners. Similar selective attack of this type can also occur with copper–aluminium, copper–tin and copper–nickel alloys but in atmospheric exposure the incidence is much less common than in the dezincification of brasses.

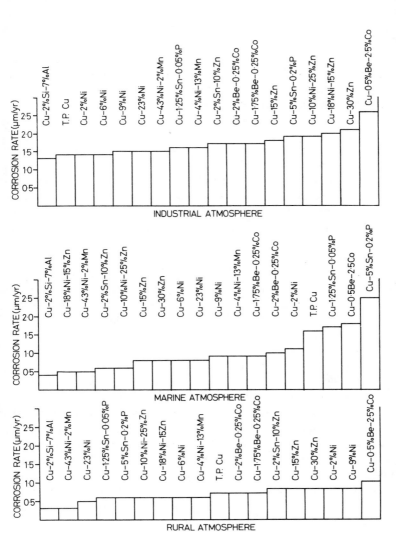

FIG. 3. Average corrosion rates for copper alloys during seven years' atmospheric exposure. (Abstracted from Ref. 63.)

A form of stress corrosion, sometimes referred to as season cracking, can occur in copper alloys. The active agent causing this type of corrosion is often ammonia which needs to be present in the environment only in minute amount in order to cause rapid stress corrosion failure when sufficient stress levels are involved either as applied design stresses in the article or internal stress generated during fabrication of the material. Fiegna *et al.*[59] studied the corrosion products produced on copper in humid atmospheres containing ammonia and found that an orthorhombic form of copper hydroxide with an unusual growth pattern was formed under these conditions, but the work does not seem to have been followed up with respect to any significance of the presence of this material in regard to the incidence of season cracking. It is well known that ammonia is frequently present in the atmosphere, being readily derived from organic matter, and the best safeguard against the occurrence of season cracking lies in the limitation of applied stresses and the elimination of high internal stresses by the use of stress-relief annealing treatments following fabrication of the material.

4.6.1 Protection of Copper

The attractive colours of copper and copper alloys and the high degree of surface finish which can be readily applied make these materials highly acceptable for decorative applications. However, it is necessary to provide some form of protection against atmospheric tarnishing if decorative finishes are to be retained for sufficiently long periods to exploit them commercially.

The use of clear lacquers as protective coatings to preserve initial finishes on copper and copper alloys is subject to the same problems of yellowing, opacity and tarnishing by acidic breakdown products of the lacquers as has been described earlier in this chapter when discussing the use of clear lacquers for the protection of zinc and its alloys. Suitable acrylic and polyurethane lacquer formulations incorporating ultra-violet absorbers, optical bleaches and inhibitive complexants for copper have been developed for use with copper alloys.[47] In these lacquers the copper sequestrant which is used in benzotriazole. Acceptable performance in excess of five years has been obtained using this type of formulation on copper and copper alloys exposed in the UK.

4.6.2 Protective Coatings of Copper

Copper and its alloys may be used as protective metal coatings or as an undercoat component in protective nickel plus chromium coating systems (see Section 4.10). When used alone as protective coatings they are generally applied to steel, either by electrodeposition or, in the case of pure copper sometimes by simple or autocatalytic chemical immersion processes. Protection is obtained as a result of the inherently greater corrosion resistance of copper than that of steel, but it is important to remember that as a result of the bimetallic effect between copper and steel the steel substrate will be preferentially attacked at any points where the coating is sufficiently defective so as to expose the steel substrate.

4.7 Tin and Tin Alloys

Tin is only very slowly corroded when exposed to atmospheric environments. The attack is uniform and only slightly reduced as the exposure period is increased. The surface of the metal is dulled and a compact layer of grey corrosion product consisting mainly of stannous oxide is formed. Hiers and Minarcik[64] recorded average corrosion rates of 1.3–1.8 μm per year in industrial exposure, 1.8–2.8 μm per year in marine exposure and 0.5 μm per year in rural exposure during a 20 year test programme in the USA.

Atmospheric pollutants have little effect on the nature of the corrosion though heavy concentrations of moist sulphur dioxide will produce a sulphide tarnish and chlorides can accelerate attack with the production of white oxychloride corrosion products.[34] Britton and Clarke[65] found that some impurities in the metal, notably zinc, may cause tarnishing in atmospheres containing sulphur dioxide.

Tin is the principal alloying constituent of many solders and when soldered joints are exposed to a corrosive environment some problems may be encountered as a consequence of the particular bimetallic couples involved, because of the particular alloy composition of the solder used, or because of the presence of soldering flux residues which may be hygroscopic. Problems of this nature are of greater consequence under immersed conditions where the corrodent has good conductivity than with the more poorly conductive electrolytes encountered in atmospheric exposure.

4.7.1 Protective Coatings of Tin

Tin may be applied as a protective coating metal for steel or for copper either by means of hot-dipping or electrodeposition processes. Because of the very high cost tin coatings are always very thin, being in the range 12–50 μm thick. At these levels of thickness porosity in the coatings can be a severe problem since the relationship between coating and substrate may be anodic or cathodic according to the particular environmental circumstances. Thus, some rusting of steel exposed at pores in tin coatings usually occurs during exposure to the atmosphere, being greater in rural or marine environments than in sulphur-polluted industrial environments. Attack at pores is enhanced under conditions of long periods of wetness.[66]

Alloys of tin such as tin–nickel and tin–lead are also used as protective coatings for steel or copper, being applied by electrodeposition processes. Tin–nickel may be deposited as a bright decorative coating which offers good protection but the deposit, which has a pinkish coloration, tends to darken on exposure and may develop cracks during service resulting in corrosion of the substrate in the cracked regions. Tin–lead electrodeposits, known as terne-plate, offer a very high degree of resistance to atmospheric corrosion and have given long service lives.

4.8 Lead and Lead Alloys

Lead and lead alloys offer a very high degree of corrosion resistance to industrially polluted atmospheres as a result of the ready formation of insoluble sulphates, sulphides, carbonates and oxides which adhere well to the metal surface and stifle further corrosive action. In chloride contaminated environments, however, the corrosion resistance is impaired by the formation of lead chloride which is more readily soluble. Tranter[67] has studied the early stages of the formation of corrosion products on lead exposed to the atmosphere in an urban area and also the patina developed on long-term exposure to urban and rural environments. His results suggest that basic lead carbonates are the major corrosion products produced in the early stages of exposure. In the intermediate stage of development of the stable patina the carbonates react with sulphur pollutants to produce a mixture of lead sulphite and lead sulphate. Finally, the lead sulphite is displaced to leave a protective film of lead sulphate.

4.8.1 Protective Coatings of Lead

Although lead is used in the form of rolled sheet for applications involving exposure to outdoor environments its cost, weight, low strength and high creep rate somewhat limit its usage. These disadvantages may be overcome by using lead in the form of a coating material applied to steel so as to utilize the high corrosion resistance of the lead coupled with the reduced weight and increased strength of the steel core material.

Coatings of pure lead and alloys of lead–tin or lead–antimony may be applied to steel by hot-dipping techniques but the process is difficult to control to ensure adequate wetting of the steel surface in the molten bath. Coatings tend to be very thin, of the order of 12–25 μm thick, and consequently pinhole porosity present in the coatings leads to early failure through rusting of steel exposed at these defects. Smith[68] has reported the results of industrial and marine exposures in the UK of steel hot-dip coated with lead–tin and lead–antimony alloys. Although the intrinsic corrosion resistance of the coatings *per se* was generally of a high order the lead–antimony alloys tended to be less resistant to chloride environments than the lead–tin alloys. Most of the test panels failed by rusting of the steel as a result of pinholing of the coatings. The incidence of pinholing was reduced by increasing the coating thickness and this led to a corresponding increase in the protective life. Lead–antimony coatings were less susceptible to pinholing than lead–tin coatings and with the lead–tin coatings the amount of pinholing was reduced by increasing the tin content of the alloy.

Lead alloy coatings may also be applied by electrodeposition and when using this method of coating pinholing is prevented. Some lead–tin alloy coatings on steel, produced by electrodeposition, were tested by Smith alongside their hot-dipped counterparts[68] and gave generally better performance.

Lead coatings may also be applied to steel by roll-bonding processes. Coatings produced by this method are much thicker than those applied by hot-dipping or electrodeposition and the working process eliminates coating porosity. This method of producing lead-coated steel was developed by the BNF Metals Technology Centre and is of fairly recent development. Long-term performance data are not yet available but preliminary atmospheric exposure tests in a UK industrial environment showed no rusting of the steel for substantial periods and no evidence of corrosion originating on steel exposed at cut edges or at the position of holes made for fasteners extending from the region of exposure to undercut the lead coating.

4.9 Nickel and Nickel Alloys

Nickel is highly resistant to atmospheric corrosion though tarnishing occurs on exposure to atmospheres contaminated with sulphur dioxide when the relative humidity exceeds 70%.[69] Heavy layers of corrosion products are not produced under free exposure conditions outdoors but such deposits may be formed on exposed surfaces which are sheltered from the washing effects of rain.

Average penetration rates for pure nickel exposed for seven years in the USA ranged from 0.25 μm per year in rural and marine environments up to 1.75 μm per year in an industrial environment. The corresponding figures obtained for nickel–iron–chromium and nickel–copper alloys were all in the range 0.02–0.75 μm per year. The corrosion rate was linear for pure nickel but decreased with increasing time of exposure for the alloys.[70] When nickel and similar nickel alloys were exposed at an industrial site in the UK, however, the corrosion rate observed was several times greater than in the industrial environment in the USA.[71]

The general effect of alloying additions on the corrosion rate of nickel in an industrial environment has been described;[70] additions of copper alone appear to have little effect but if small amounts of iron are added to the nickel–copper alloys a marked improvement in corrosion resistance is achieved. However, larger amounts of added iron—e.g. 20% or more—markedly increase the corrosion rate of the nickel–copper alloys.

Since nickel is so resistant to atmospheric corrosion it is very widely used as an electrodeposited coating for the protection of ferrous and non-ferrous metals, in which role it performs very efficiently. However, because of the tarnishing which rapidly develops in outdoor exposure it is usually over-coated with thin deposits of chromium or of one of the precious metals in order to retain a bright decorative appearance. When this is done any discontinuities in the topcoat metal provide sites where increased corrosion of the nickel occurs, due to the bimetallic effect, and localized pitting corrosion of the nickel layer ensues. Furthermore, when the corrosion pits penetrate through the nickel layer to expose the substrate metal the additional bimetallic effects introduced cause accelerated attack on the substrate metal, both in the case of ferrous and of non-ferrous substrates.

4.10 Nickel Plus Chromium Decorative and Protective Coatings

Electrodeposited coatings of nickel plus chromium are applied to substrates of steel, zinc and its alloys, copper and its alloys and to aluminium and its alloys to provide a protective coating system which is capable of retaining a good decorative appearance for considerable periods of exposure to atmospheric environments.

The composition and physical condition of the components of the coating system and the thicknesses of the various coating layers all influence the corrosion performance as also does the nature of the environment to which the system is exposed.

The thin electrodeposited chromium topcoat always contains defects and discontinuities which expose minute areas of the underlying nickel coating to the corrodent electrolyte. Localized pitting corrosion develops at these sites with eventual penetration to any undercoats and to the substrate metal. Consequently the first requirement for adequate protective performance is a sufficient thickness of nickel coating.

In order to achieve maximum brightness and levelling of the coating system nickel is usually deposited in the 'bright' form by using electroplating baths which contain brightener and levelling agent additives. These additives increase the sulphur content of the nickel deposit and these bright deposits have a greater corrosion pitting penetration rate than the sulphur-free or low sulphur 'dull' nickels. In order to improve corrosion performance duplex nickel systems have been developed in which that portion of the nickel adjacent to the undercoat or substrate is of a low-sulphur dull or semi-bright nickel and the outer portion of the deposit is of the high-sulphur bright nickel. When corrosion pitting commences in the bright nickel portion of the duplex deposit at the discontinuities in the chromium topcoat, penetration of the bright layer continues until the semi-bright layer is exposed. The bright nickel then provides some sacrificial protection to the semi-bright layer causing lateral spread of the pits in the bright layer with a reduced penetration rate through the semi-bright layer and consequently increasing the period of protection of the substrate metal (see Fig. 4).

When pitting corrosion completely penetrates the nickel layer to expose the undercoat or substrate metal the coating system is cathodic to the undercoat or substrate and accelerated corrosion occurs preferentially with undercutting of the coating layer (see Fig. 5). Furthermore, the corrosion products produced by this preferential attack are more disfiguring than those produced by corrosion of the nickel layer and the adverse effect on decorative appearance becomes excessive.

Conventional chromium deposits—known as 'regular' chromium—used as topcoats are normally of 0.3 μm minimum thickness and always contain the pores and discontinuities which provide the sites for the pitting corrosion of the nickel layer. Increasing the thickness of the chromium deposit may reduce but will not completely eliminate discontinuities and the smaller number of defects pit at a faster rate because of the adverse anode/cathode ratio; furthermore, as chromium thickness is

FIG. 4. Lateral spread of corrosion in bright nickel layer of duplex nickel electrodeposit. (Magnification × 750.) Reproduced by kind permission of BNF Metals Technology Centre.

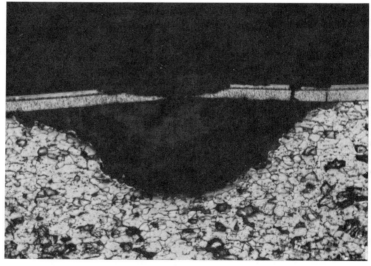

FIG. 5. Preferential corrosion of steel substrate at pit penetrating through nickel and chromium electrodeposit. (Magnification × 140.) Reproduced by kind permission of BNF Metals Technology Centre.

increased the brittle nature of the deposit induces unsightly macrocracking which itself provides additional corrosion sites. Alternatively, by deliberately increasing the number of discontinuities the anode/cathode relationship is favourably influenced and pitting rate reduced. By modification of the chromium deposition process a very làrge number of small discontinuities, invisible to the naked eye and having no adverse effect on the bright reflective properties of the chromium, can be produced. These deposits, known as 'microdiscontinuous' chromium, enable greatly improved protection of the substrate to be achieved with much longer acceptable service lives. Two types of microdiscontinuous chromium may be produced:

1. *Microcracked chromium* having an invisible network of cracks forming a closed pattern with more than 250 cracks per 10 mm over the whole surface and a minimum thickness of 0.8 μm (microcracked chromium may also be satisfactorily produced at 0.3 μm thickness by the deposition of regular chromium on a thin layer of highly stressed nickel itself deposited on the protective nickel layer). See Fig. 6.

2. *Microporous chromium* having a minimum of 10 000 micropores per 100 mm^2 over the whole surface and a minimum thickness of 0.3 μm. The invisible micropores are produced by deposition of regular chromium on to a layer of non-conducting particles co-deposited in the surface of the protective nickel layer. Alternatively, chromium deposited from a trivalent chromium bath is always produced in the microporous condition.[72]

When corrosion of the nickel exposed at microdiscontinuities occurs the attack is more superficial than under regular chromium deposits giving longer periods of protection but in some severe atmospheric environments such as severely polluted industrial atmospheres the spread of corrosion in the surface of the nickel deposit causes microflaking of the chromium, with the consequent development of surface dulling over long periods of exposure.[7,73]

When copper undercoats are used beneath nickel deposits and penetration of the nickel occurs, the copper corrosion products produced are of a particularly disfiguring nature as they spread on the surface of the coated article. There is dispute, therefore, as to the benefits or disadvantages of the use of copper undercoats when rapid pitting may occur as with regular chromium deposits. However, there is considerable evidence that when microdiscontinuous chromium deposits are used copper

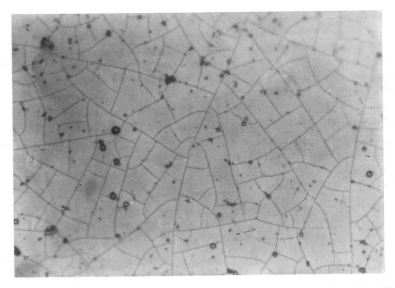

FIG. 6. Microcrack pattern in chromium electrodeposit. (Magnification × 1000.) Reproduced by kind permission of BNF Metals Technology Centre.

undercoats can improve considerably the overall corrosion performance of the coating system.[7,73-75]

The large number of combinations of individual components of the nickel plus chromium system is reflected in the minimum thickness requirements laid down for service use in outdoor environments in BS 1224:1970.[76] Nickel thicknesses range from 20 to 40 μm minimum according to the type of nickel and chromium employed, the nature of the substrate and the presence or absence of copper undercoats.

There has been very extensive study of the performance of the various nickel plus chromium coating systems when exposed to the atmosphere and copious data are available.[7,72-75,77-82] As might be expected the severest and most rapid corrosion occurs in industrial atmospheres polluted with sulphur gases where the period of protection of the substrate may be as little as three months or less for the poorer coating systems and as much as two years or more with the best systems. In exposure to marine environments where sulphur pollution is absent the life expectancy may be increased three or four times and in completely unpolluted rural atmospheres several years' life expectancy is readily achieved with all but the poorest coating systems. The incidence and extent of surface dulling with microdiscontinuous chromium systems is also dependent on the degree of pollution of the environment, being greatest and most rapidly produced where static exposure to industrial environments is involved. In these conditions an unacceptable appearance due to surface dulling may develop in one to two years. However, surface dulling is very substantially delayed and reduced in effect when the exposure is under mobile conditions as in service on motor vehicles—one of the principal uses of this type of coating system—so that it often does not have any significant adverse effect on appearance over a period of several years. For this reason it is important to supplement static exposure testing with mobile exposure when evaluating the performance of systems intended for this type of service.

REFERENCES

1. British Standard BS 5493:1977. British Standards Institution.
2. ISO/DIS 4542. International Standards Organization.
3. V. E. Carter and H. S. Campbell. ASTM STP 435 (1968). Amer. Soc. Test. Mat.
4. V. E. Carter. *Br. Corros. J.*, **9** (1), 10 (1974).
5. British Standard BS 1615:1972. Appendix W. British Standards Institution.
6. British Standard BS 3987:1974. Appendix F. British Standards Institution.
7. V. E. Carter. *Trans. Inst. Met. Fin.* **48** (1), 19 (1970).
8. British Standard BS 5466:Pt I:1977. British Standards Institution.
9. British Standard BS 5466:Pt II:1977. British Standards Institution.
10. British Standard BS 5466:Pt III:1977. British Standards Institution.
11. British Standard BS:5466:Pt IV:1979. British Standards Institution.
12. British Standard BS 1615:1972. Appendix H. British Standards Institution.
13. British Standard BS 5466:Pt V:1979. British Standards Institution.
14. J. Edwards and V. E. Carter. *Trans. Inst. Met. Fin.*, **40** (2), 48 (1963).
15. V. E. Carter. *Trans. Inst. Met. Fin.*, **45** (2), 64 (1967).
16. ASTM Test Method. A164–55. Amer. Soc. Test. Mat.
17. ASTM Test Method. A165–55. Amer. Soc. Test. Mat.
18. F. L. McGeary *et al.* ASTM STP 435, 141 (1968). Amer. Soc. Test. Mat.
19. H. Sutton. *Welding Met. Fab.*, **45** (3), 163 (1977).
20. W. A. Bell and H. S. Campbell. *J. Inst. Met.*, **89**, 464 (1960–61).
21. S. Lindgren *et al.*, *Proc. 7th Scand. Corr. Congress*, 1975.
22. V. D. Kalinin. *Prot. Metals*, **12** (5), 501 (1976).
23. L. Atteras and D. A. Hagerup., *Proc. 7th Scand. Corr. Congress*, 1975.
24. F. Pearlstein and L. Teitell, *Mats. Perform.*, **13** (3), 22 (1974).
25. Yu. N. Mikhailovskii. *Prot. Metals*, **9** (3), 240 (1973).
26. H. Sick. *Aluminium* (Dusseldorf), **54** (4), 268 (1978).
27. P. Doig and J. W. Edington. *Br. Corr. J.*, **9** (4), 220 (1974).
28. J. M. Truscott *et al.*, *J. Inst. Met.*, **99**, 57 (1971).

29. V. E. Carter and H. S. Campbell, *J. Inst. Met.*, **89**, 472 (1960–61).
30. V. E. Carter and H. S. Campbell, *Br. Corr. J.*, **4** (1), 15 (1969).
31. V. E. Carter, *J. Inst. Met.*, **100**, 208 (1972).
32. V. E. Carter, *Br. Corr. J.*, **9** (1), 10 (1974); *Aluminium*, **49**, 682 (1973).
33. E. Wettinck, *Aluminium*, **52** (2), 727 (1976).
34. L. L. Shrier, *Corrosion*, 2nd edn 1976. Newnes-Butterworth.
35. B. N. Rybakov *et al.*, *Sov. J. Non-Ferr. Met.*, **16** (2), 64 (1975).
36. S. M. Brandt and L. H. Adam, ASTM STP 435, 95 (1968). Amer. Soc. Test. Mat.
37. M. L. Greenlee and L. F. Plock, ASTM STP 435, 33 (1968). Amer. Soc. Test. Mat.
38. J. C. Hudson and J. F. Stanners, *J. Appl. Chem.*, **3**, 86 (1953).
39. H. Guttman, ASTM STP 435, 223 (1968). Amer. Soc. Test. Mat.
40. G. Schickorr, *Metall.*, **15**, 981 (1961).
41. C. J. Slunder and W. K. Boyd. *Zinc: its corrosion resistance*, 1971, Zinc Development Assoc.
42. G. Schickorr, *Werkstoffe und Korrosion*, **15**, 537 (1964).
43. B. Sanyal *et al.*, *J. Sci. & Ind. Res.* (India), **15B**, 448 (1956); **18A**, 127 (1959); **20D**, 27 (1961); **21D**, 185 (1962).
44. P. T. Gilbert, *J. Appl. Chem.*, **3**. 174 (1953).
45. S. R. Dunbar, ASTM STP 435, 308 (1968). Amer. Soc. Test. Mat.
46. Report of Sub-Committee V of ASTM Committee B-6. 1961. *Proc. Amer. Soc. Test. Mat.*, **61**, 273.
47. I. R. A. Christie and V. E. Carter, *Trans. Inst. Met. Fin.*, **50** (1), 19 (1972).
48. C. L. Hippensteel and C. W. Borgmann, *Trans. Amer. Electrochem. Soc.*, **58**, 23 (1930).
49. T. Biestek, *Met. Finishing*, **68** (4), 48 (1970).
50. W. Radeker *et al.*, *Proc 6th Int. Conf. on Hot Dip Galv.*, Interlaken. 1961. p. 238. Zinc Development Assoc.
51. S. E. Hadden, *J. Iron & Steel Inst.*, **171**, 121 (1952).
52. H. S. Campbell *et al.*, *J. Iron & Steel Inst.*, **203**, 248 (1965).
53. V. E. Carter, *Met. Fin. J.*, 304 (1972).
54. D. N. Layton, *Trans. Inst. Met. Fin.*, **43** (4), 153 (1965).
55. E. Mattsson and R. Holm, ASTM STP 435, 187 (1968). Amer. Soc. Test. Mat.
56. W. H. J. Vernon, *J. Inst. Met.*, **49**, 153 (1932).
57. J. R. Freeman and P. H. Kirby, *Metals & Alloys*, **5**, 67 (1934).
58. E. Niskanen and U. M. Franklin, *Canad. Metall. Quart.* **9** (1), 339 (1970).
59. A. Fiegna *et al.*, *Corros. Sci.*, **12** (8), 673 (1972).
60. R. Ericsson and T. Sydberger, *Werkstoffe und Korrosion*, **28** (11), 755 (1977).
61. E. Mattsson and R. Holm, *C. D. A. Journal 'Copper'*, No 27. Spring 1966. Copper Development Assoc.
62. W. H. J. Vernon, *Trans. Faraday Soc.*, **27**, 255 and 582 (1931).
63. D. H. Thompson, ASTM STP 435, 129 (1968). Amer. Soc. Test. Mat.
64. G. O. Hiers and E. J. Minarcik, ASTM STP 175, 135 (1956). Amer. Soc. Test. Mat.
65. S. C. Britton and M. Clarke, *Trans. Inst. Met. Fin.*, **42** (5), 195 (1964).
66. S. C. Britton, *The corrosion resistance of tin and tin alloys*, 1951. Tin Research Institute.

67. G. C. Tranter, *Br. Corr. J.*, **11** (4), 222 (1976).
68. R. Smith, *Sheet Met. Ind.*, **49** (12), 761 (1972).
69. W. H. J. Vernon, *J. Inst. Met.*, **48**, 121 (1932).
70. D. van Rooyen and H. R. Copson, ASTM STP 435, 175 (1968). Amer. Soc. Test. Mat.
71. T. E. Evans, *Proc 4th Int. Cong. on Met. Corr.*, 1969, p. 408.
72. C. Barnes *et al.*, *Trans. Inst. Met. Fin.*, **55** (2), 73 (1977).
73. V. E. Carter, *Trans. Inst. Met. Fin.*, **48** (1), 16 (1970).
74. N. W. Phasey, *Trans. Inst. Met. Fin.*, **51** (2), 77 (1973).
75. R. J. Clauss and R. W. Klein, *Proc 7th Int. Met. Fin. Conf.*, Hanover, 1968 (Interfinish 68), p. 124.
76. British Standard BS 1224:1970. British Standards Institution.
77. P. C. Crouch and A. C. Hart, *Trans. Inst. Met. Fin.*, **52** (2), 59 (1974).
78. J. K. Dennis and P. Tipping, *Trans. Inst. Met. Fin.*, **52** (1), 5 (1974).
79. V. E. Carter and I. R. A. Christie, *Trans. Inst. Met. Fin.*, **51** (2), 41 (1973).
80. J. K. Dennis and J. J. Fuggle, *Trans. Inst. Met. Fin.*, **49** (2), 54 (1971).
81. R. L. Saur, *Plating & Surface Finishing*, **65** (3), 48 (1978).
82. A. H. Du Rose, *Plating*, **62** (10), 941 (1975).

Chapter 3

ASPECTS OF MICROBIAL CORROSION

A. K. TILLER

National Corrosion Service, National Physical Laboratory, UK

1. INTRODUCTION

Microbial corrosion is only one of many forms of corrosion which contribute to substantial losses to the economy and pollution of the environment. Its study is an interdisciplinary subject which requires at least some understanding in the fields of chemistry, metallurgy, microbiology, and biochemistry. By definition it refers to the degradation of metallic structures resulting from the activity of a variety of organisms which either produce aggressive metabolites to render the environment corrosive, or are able to participate directly in the electrochemical reactions occurring on the metal surface. In many cases microbial corrosion is closely associated with biofouling phenomena, which are caused by the activity of organisms that produce deposits of gelatinous slime or biogenically-induced corrosion debris in aqueous systems. Familiar examples of this problem are the growth of algae in cooling towers and barnacles, mussels and seaweed on marine structures. These growths either produce aggressive metabolites or create microhabitats suitable for the proliferation of other bacterial species, e.g. anaerobic conditions favouring the well known sulphate-reducing bacteria. In addition, the presence of growths or deposits on a metal surface encourages the formation of differential aeration or concentration cells between the deposit and the surrounding environment, which may stimulate existing corrosion processes.[1,2]

However, although microbial corrosion is frequently associated with biofouling, the area in which it causes most widespread concern is the deterioration of iron and steel structures buried in soils and sediments.

The seriousness of this problem in the United Kingdom has been confirmed by a recent survey,[3] which also highlighted the fact that the continued incidence of microbial corrosion could be attributed to a general lack of awareness of the problem. This lack of awareness is somewhat surprising since there has been a wealth of published information available for some time.[4] There has been a tendency in some of the literature of the past decade to suggest that microbial corrosion is a newly discovered problem. This is not so; the scientific study of the subject began at least 70 years ago, and the purpose of this review is to examine its historical development and the contribution of a variety of research investigations to an understanding of the mechanisms by which micro-organisms are responsible for some of the more serious problems encountered in industry. The methods available for assessing the corrosivity of environments, particularly soil environments, and the methods available for the prevention and control of microbial corrosion will also be considered.

The economic impact of microbial corrosion is very significant. In 1968 Knox[8] estimated that biodeterioration of materials was costing at least $1000 million per year, to which a major contribution was made by microbial corrosion. Several estimates have been made of the cost of microbial corrosion alone. In 1972 Iverson[6] reported that the cost of microbial corrosion of buried pipelines to the USA was approximately $500 million to $2000 million per year and that in the petroleum industry, for example, 77% of the failures in one group of oil-producing wells were attributed to this phenomenon.

A number of estimates have been made in the United Kingdom over the years which have ranged from £20 million per year for the repair and maintenance of buried pipelines[9] to £300 million per year for problems arising from infections of industrial cooling water systems.[10] In a recent detailed survey of the problems of microbial corrosion in the UK Le Roux and Wakerley[11] were unable to quote specific costs for any of the industries investigated. They concluded that there were insufficient data available to attribute costs accurately for such a specific problem, and concurred with comments from some of the large organizations surveyed on the difficulty of separating the cost for this single facet from that for corrosion as a whole. However major pipeline users predicted that losses due to microbial problems should decrease markedly over the next decade because of the improved protective systems which are now being applied. In addition, the situation should improve because greater emphasis is being laid on the control of potential problems by implementing specifications and codes of practice which are available from a variety of sources.

A survey of the industrial cost of biofouling[12] indicates that the current cost of this problem is at least £300–500 million per year. This includes estimates of the cost of lost production, maintenance, increased fuel consumption, and increased capital equipment requirements. However, although some of the major industries recognize the implications of these problems and the value of good housekeeping,[11] a large number of the smaller organizations are either less aware of the problems of microbial corrosion or adopt relatively cheap remedial measures to reduce maintenance costs, in many cases resulting in the premature breakdown of plant and equipment. Clearly therefore it is important to emphasize the benefits of good housekeeping and a vigilant approach to remedial procedures, but before these are widely accepted, it seems still to be necessary in many instances to convince engineers (and their accountants) of the reality of microbiological factors in corrosion problems.

2. HISTORICAL DEVELOPMENT

The phenomenon of microbial corrosion was reported before the turn of the century when Garrett[13] postulated that increased corrosion of a lead-covered cable could be attributed to the metabolites of bacterial activity. In this case he referred to the formation of ammonia, nitrites and nitrates. By 1910 Gaines[14] had defined the problem more clearly, producing evidence which clearly indicated that iron bacteria and sulphur bacteria were responsible in part for the corrosion of buried ferrous metals. In many instances the corrosion was also associated with the presence of oxygen.

However, in 1934 von Wolzogen Kühr and van der Vlugt[15] reported on the corrosion of ferrous metal buried in an anaerobic clay soil. Their studies indicated that corrosion had been caused by the activity of the obligate anaerobic bacteria known as the sulphate reducers. They proposed that the organisms played a direct role in the corrosion mechanism, utilizing cathodic hydrogen and thus stimulating the anodic dissolution of iron.

Prior to this work other workers had reported the formation of deposits in water pipes by iron bacteria, in particular the aerobic species *Sphaerotilus natans*, *Caulobacter* and *Gallionella*.[16,17] Evidence for corrosion by a variety of both anaerobic and aerobic micro-organisms continued to be published.[18-22] In 1953, Uhlig[23] reported on the role of organisms such as slime-forming bacteria, fungi, algae, protozoa and diatoms, and commented that their primary contribution to the corrosion process consisted of the formation and maintenance of differential aeration and concentration cells. In 1971 the extensive problems of corrosion of machinery components and finished articles was well documented by Hills,[24] the mechanisms in this instance being attributed to the formation of aggressive metabolites such as organic acids. In nearly all cases corrosion occurred at the solution/metal/air interface where there was a plentiful supply of oxygen, water and acid. It was during the same period that the corrosion of aluminium alloy fuel tanks was reviewed,[25] the problem being attributed to the action of particular fungi and bacteria. Recent studies[26] have confirmed that the predominant factor is the activity of the fungi *Cladosporium resinae* and *Pseudomonas aeruginosa* leading to the formation of organic acids, and that the influence of bacteria in this instance is slight.

In addition to the case histories reported for metals in contact with soils and waters, evidence had been collected which clearly indicated that similar degradation problems could occur on non-metallic structures, in particular on concrete.[27-30] The main organisms involved in this case are the sulphur-oxidizing bacteria, and Pochon and Jaton[31] reported several instances of this type of biodeterioration which he attributed to the bacterial oxidation of atmospheric sulphur pollutants on the surfaces of the structure.

Several reviews on the subject of microbial corrosion have been published since 1971.[4-7] The importance of this aspect of corrosion has increased considerably with the advent of the energy crisis. The problems of biofouling in industrial cooling systems and marine fouling on offshore platforms, together with those of sulphide contamination of oil and gas reservoirs, have emphasized the need for a better and more widespread understanding of the role which bacteria play in the corrosion of metals. It is particularly significant that in the majority of reported case histories the role of the sulphate-reducing bacteria is mentioned. These organisms, of the genera *Desulfovibrio* and *Desulfotomaculum*, are without doubt the most troublesome micro-organisms involved in the corrosion of iron and steel. Corrosion failures attributable to their activity, and resulting in costly down-time and remedial work, have been experienced by many industrial organizations,[3] although in many cases the involvement of micro-organisms has not been recognized, either because of a lack of awareness or of inexperience in the identification of this type of problem.

3. PHENOMENOLOGY OF MICROBIAL CORROSION

It is a reasonable assumption that corrosion is a potential problem in any environment in which micro-organisms thrive. However, the activity of specific species may be limited by the environment and conditions favourable for the proliferation of one species may be quite inimical to another. Because observation is complicated by the presence of conventional corrosion products, fouling and other deposits, detection of microbial corrosion by simple visual examination is usually difficult, and it is even more complex to attribute the correct proportion of responsibility when microbiological activity is only one of a number of causative factors. However there are a number of specific phenomenological aspects of the morphology of the attacked surface which can be used in many cases to indicate the organism responsible.

In aqueous environments, biofouling by algae, microbial slime masses, barnacles, mussels, or seaweed provides potential microhabitats which may allow proliferation of other bacterial species whose activity results in localized corrosion beneath the fouling. When the metal surface is examined, the pits may be filled with a soft black product with a strong smell of sulphide, beneath which the pitted region appears as bright metal. This is typical of corrosion caused by sulphate-reducing bacteria.[5] In fresh water transmission pipelines the first indications of a problem may be excessive tuberculation on the inside of the pipe,[32] indicating the presence of the iron-oxidizing bacteria. The shell of the tubercle usually consists of a hard crust of yellow-brown ferric hydroxide, and again pitting is often observed beneath the tubercle; in some cases the typical soft black sulphide product with a clean, bright-metal pitted region described above suggests the activity of sulphate-reducing bacteria within the habitat generated by the iron oxidizers.

The corrosion of aluminium aircraft fuel tanks is associated with fungal growth in the presence of condensed water. The colour of the growth may be indicative of the organism, for example *Cladosporium resinae* often has a pinkish tone.[7] During colonization of the surface by this fungus, pH changes usually occur, and with this type of corrosion values between 3 and 4 are common. When the fungal growths are removed pitting corrosion is usually observed.

Corrosion which occurs during machining and fabrication operations can frequently be attributed to biodeterioration of coolants or lubricants. Emulsions used in machine operations are usually chemically stabilized dispersions of oil in water, stability being achieved by the use of anionic or non-ionic surfactants as emulsifying agents. When freshly prepared these are mildly alkaline materials which favour the growth of numerous aerobic organisms. The first indication of the problem is usually a change from an alkaline to an acid pH but a more obvious diagnostic sign is often by the pungent smell of hydrogen sulphide when the line has been stagnant during a weekend period. Another common symptom is the 'creaming' of the fluid resulting in the separation of free oil from the emulsion. If this is allowed to progress a slurry of bacterial slime develops. Hills[24] developed an electrical capacitance method for monitoring the stability of emulsion systems as well as a simple testing kit which allows the bacterial population to be monitored. Where synthetic coolants are used, the first indication of biodeterioration is a strong smell of ammonia resulting from the degradation of the sodium nitrite additive. This also produces a marked increase in pH to alkaline values in excess of pH 10. When this type of degradation occurs where copper and its alloys are being machined, green staining of the metal results.

The bacteria of the genus *Thiobacillus* are fairly common organisms capable of oxidizing sulphur and reduced sulphur compounds to free sulphuric acid. The first symptoms of the activity of these organisms is a marked increase in the acidity of the environment, usually either soil or deposits produced by atmospheric pollution. In the case of soils the marked change in pH indicates a ready supply of sulphur compounds, and it is not unusual for the organisms to be oxidizing hydrogen sulphide which is generated by sulphate-reducing bacteria that are active in lower soil strata. The slow diffusion of hydrogen sulphide through the soil is the primary cause of the corrosion that occurs under these conditions, the area attacked usually having an even etched appearance with local pH values in the range 2–4 common. The same type of organism is also responsible for the degradation of concrete sewers and other non-metallic structures; details of the mechanism by which this occurs are discussed below later.

As indicated above the most generally significant microbial corrosion problems are those associated with buried metals, particularly where anaerobic waterlogged conditions prevail. Such situations are usually associated with clays or clayey soils, but severe anaerobic corrosion of metallic structures has been experienced in untypical soil environments where external influences have promoted conditions conducive to proliferation of the sulphate-reducing bacteria. The corrosion problems encountered under these conditions are again easily recognized either by the smell of hydrogen sulphide when the site is excavated or the presence of black stained soil in the region of the buried structure. The presence of black films of iron sulphide and the strong smell of hydrogen sulphide when either the soil or the corrosion product is treated with dilute hydrochloric acid also distinguish this type of corrosion. The sulphide film is soft and loose and when the film is removed the underlying pitted region is initially of bright metal which flash rusts rapidly. In some cases the presence of a yellowish white deposit of elemental sulphur is observed around the periphery of the pits. A type of corrosion peculiar to cast iron is graphitic corrosion, or more colloquially graphitization; in certain areas the iron is dissolved from the metal leaving a mass of graphite and corrosion product almost indistinguishable to the eye from the rest of the pipe, but soft enough to be pierced by a sharp instrument. When these symptoms are discovered in such environments as described above it can be assumed that sulphate-reducing bacteria are involved and their presence can in fact usually be demonstrated in the corrosion product and the surrounding soil in very much larger numbers than the average background 'count' for the district.

Thus, empirical evaluation can frequently be used to recognize the involvement of microbiological agencies in the corrosion process. The indications may be summarized into three broad categories.

1. Conditions conducive to microbiological activity either in the general environment or in microenvironments generated by biofouling or other means. These should include pH values generally between 4 and 9 (although sulphur oxidizers will tolerate lower levels); temperatures generally below 50–60 °C (although some thermophilic strains will tolerate up to 70 °C); appropriate redox conditions and an appropriate nutrient source for activity of the organisms considered.

2. Typical morphology and/or chemistry of corrosion products, for example ferric hydroxide tubercles, soft black sulphide products in pits, graphitized areas in cast iron structures.

3. Typical changes in the environment resulting from microbial activity, for example marked changes in pH in either acid or alkaline direction, high levels of sulphide in soils adjacent to structures.

Obviously several such indicators should be sought before it is reasonable to assume that microbial corrosion has occurred, and in many cases further testing will be necessary to justify this diagnosis, although some types of problem are sufficiently common to be interpreted on the basis of experience alone.

4. MICRO-ORGANISMS ASSOCIATED WITH CORROSION

Micro-organisms from a wide range of genera and species are encountered in corrosion problems. They may be classified in three general groups: (i) fungi; (ii) algae and diatoms; and (iii) aerobic and anaerobic bacteria. In order to understand how these organisms may be associated with the corrosion of metals it is helpful to review briefly some of their more important characteristic features.

4.1 Fungi
The fungi are a group of organisms which grow as saprophytes on nonliving organic matter, or as parasites on other living organisms. They are extremely numerous and classification is often a problem. Those associated with corroding metal are generally filamentous or yeast-like. The filaments are branched and form a tangled mass referred to as mycelium. In contrast to some bacteria, fungi require oxygen for growth. Energy for growth is obtained either by anaerobic fermentation or aerobic oxidation of an organic substrate, which releases organic acid metabolites as the end product, e.g. citric acid, oxalic acid. Since they can utilize hydrocarbons as a source of carbon they may be active in largely hydrocarbon environments, and have been the causative agents in the corrosion of aluminium aircraft fuel tanks.

4.2 Algae and Diatoms
The algae are a heterogeneous group of chlorophyll-containing plants found in sea and fresh waters. They are primarily autotrophic, obtaining their energy from light or by the oxidation of inorganic materials and their carbon by the assimilation of carbon dioxide. These organisms are commonly found in cooling water systems. Their role is similar to that of the fungi; by adhering to the metal surface they encourage the formation of differential aeration and concentration cells and thereby accelerate already existing corrosion processes. The diatoms are a group of unicellular algae which predominate in fresh and salt water environments as well as in soils.

4.3 Bacteria
The bacteria associated with the corrosion of metals are unicellular, possessing a thick rigid cell wall and dividing by binary fission. Some are motile by means of a flagellum. These organisms can be either autotrophic or heterotrophic and either aerobic or anaerobic. Heterotrophs

obtain both their energy and carbon requirements from organic sources and assimilate carbon dioxide to only a limited extent. Autotrophs are bacteria which obtain their energy from light or by the oxidation of inorganic materials and their carbon by assimilation. Anaerobic bacteria do not require oxygen for their growth. The unicellular bacteria have three basic shapes; rod-like, curved or spirilloid, and spherical. They vary considerably in size. Some unicellular bacteria are stalked such as the iron bacteria *Gallionella*. Other iron bacteria are filamentous, and all obtain the energy for carbon dioxide fixation by oxidation of ferrous ions to ferric ions with the consequent precipitation of ferric hydroxide. These organisms are responsible for the formation of tubercles in potable water transmission pipelines. The general physiology and taxonomy of the iron bacteria have been reviewed by Pringsheim[33] and Stokes.[34]

The most important bacteria associated with corrosion processes are those in whose metabolism sulphur and/or its compounds play a vital part. These organisms are intimately concerned with the sulphur cycle in nature. This is a continuous cycle of biochemical change, illustrated in Fig. 1. The upper part of this cycle, the oxidation of sulphur to sulphuric acid is performed by the aerobic bacteria of the genus *Thiobacillus*, usually referred to as the sulphur oxidizers. The 'short circuit' of the cycle, i.e. the reduction of sulphate to hydrogen sulphide is carried out by anaerobic bacteria of the genera *Desulfovibrio* and *Desulfotomaculum*, the

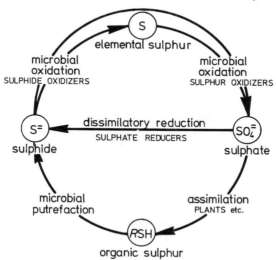

FIG. 1. The sulphur cycle.

sulphate reducers. Although the sulphur oxidizers are implicated in microbial corrosion their involvement is generally only slight compared with the problems generated by the presence of the sulphate-reducing bacteria. Before discussing the specific manner in which these organisms are involved with microbial corrosion, some of their more important and general characteristics are given below.

4.3.1 Thiobacillus spp.—The Sulphur Oxidizers

These autotrophic bacteria are short non-sporulating acidophilic rods approximately 0.5×1.0–1.5 μm in size. They are common soil organisms and usually occur either as single cells or in pairs, and are motile. The optimum temperature for their growth is 25–30 °C; above 55–60 °C they are killed. Two main species are involved in microbial corrosion, *Thiobacillus thioparus* and *Thiobacillus thiooxidans*—which include the species often referred to as *Thiobacillus concretivorus*. This latter organism is responsible for the degradation of concrete and building stones. The chemistry by which these organisms generate sulphuric acid has been discussed in detail by Booth[5] and Purkiss,[10] the generally accepted pathway[10] involving the following reactions, which may be independent or interlinked depending to the substrate and environment.

$$2 H_2S + 2O_2 \rightarrow H_2S_2O_3 + H_2O$$
$$5 Na_2S_2O_3 + 4 O_2 + H_2O \rightarrow 5 Na_2SO_4 + H_2SO_4 + 4 S$$
$$4 S + 6 O_2 + 4 H_2O \rightarrow 4 H_2SO_4$$

4.3.2 Desulfovibrio spp. and Desulfotomaculum spp.—The Sulphate-Reducing Bacteria

The expression sulphate-reducing bacteria is conventionally reserved for a class of bacteria which conducts dissimilatory sulphate reduction. In this process the sulphate ion acts as an oxidizing agent for the dissimilation of organic matter, as does oxygen in conventional respiration. A small amount of sulphide is assimilated by the organism but virtually all is released as sulphides. The biochemistry of the metabolic pathway of sulphate reduction is complex and has been reviewed by Peck[35] and Postgate.[36] It involves two or possibly three enzymes which are collectively known as the sulphate-reductase enzymes. The other important feature of these organisms is that while the reduction process is taking place they are able to take up molecular hydrogen according to the reaction:

$$SO_4^{2-} + 4 H_2 \rightarrow S^{2-} + 4 H_2O$$

This catalysis of sulphate is carried out by the enzyme hydrogenase. It must be emphasized however that the presence of this enzyme is not essential for the reduction of sulphate. The importance of this is discussed in more detail later.

These organisms are strict anaerobes although they can survive for long periods under well-aerated conditions. The two well-established genera are *Desulfovibrio* and *Desulfotomaculum*. Table 1 lists the species identified and their classification.[37] The important feature about this group of organisms is that they tolerate a wide range of environments and may appear as mesophilic, thermophilic or halophilic strains. The optimum growth rate for the mesophiles and halophiles is 25–44 °C at pH values between 5.5 and 9.0 (optimum pH 7.2). The thermophilic strain *Desulfotomaculum nigrificans* tolerates temperatures up to 70 °C, but the optimum temperature for growth is usually 55 °C. For identification purposes tests in various inhibitor solutions are often used. The *Desulfovibrio* spp. can show an exceptional resistance to inhibitors[38] while *Desulfotomaculum* is normally more sensitive.

Although many of the characteristics of these bacteria are known, unexpected features still arise and in many cases the mechanisms by which microbial corrosion is induced or exacerbated are not fully understood.

5. MECHANISMS OF MICROBIAL CORROSION

5.1 General Considerations

Before discussing the specific mechanisms involved in microbial corrosion it must be emphasized that microbial involvement in the field of corrosion does not involve any new form of corrosion process. The corrosion of metals in an aqueous environment is well recognized as an electrochemical phenomenon in which part of the metal substrate is oxidized and transferred into solution, the anodic reaction. The balancing cathodic reaction is the simultaneous reduction of some component in the corrosive environment to preserve overall electroneutrality. The metal ions in solution may subsequently be precipitated as insoluble products which may be loose and bulky or become firmly adherent to the metal surface, to which they are protective to a greater or lesser extent. In favourable circumstances protective films can develop under natural conditions, a typical example of this phenomenon being the well-preserved archaeological iron objects excavated from a corrosive soil

TABLE 1

LIST OF THE SULPHATE-REDUCING BACTERIA, WITH SOME OF THEIR MORE IMPORTANT CHARACTERISTICS (FROM MILLER AND TILLER[37])

Species	Synonyms (old names)	Characteristics	
Genus *Desulfovibrio*:	*Vibrio, Sporovibrio, Spirillum, Microspira*	Single flagellum, or a tuft. Hydrogenase usually present	Do not form spores.
Dv. *desulfuricans*	*Desulfovibrio desulfuricans*	Curved rods (vibrios); sometimes spirilloid, occasionally straight. Typical size 3–5 μm × 0.5–1 μm. Single flagellum	Growth on pyruvate or choline in sulphate-free media
Dv. *vulgaris*			Obligate salt-water species (requires Cl$^-$)
Dv. *salexigens*			
Dv. *gigas*		Large curved rods or spirilla, 5–10 μm × 1.2–1.5 μm. Tuft of flagella	
Dv. *africanus*		Long, slender, sigmoid rods, 5–10 μm × 0.5 μm. Tuft of flagella	
Genus *Desulfotomaculum*: Dt. *nigrificans*	*Vibrio thermodesulfuricans, Clostridium nigrificans*	Peritrichous flagella. Spores formed The only thermophilic species (optimum temperature 55°C). Rods, 3–6 μm × 0.3–0.5 μm. Hydrogenase activity variable; not coupled to sulphate reduction. Growth on pyruvate in sulphate-free media	
Dt. *orientis*	*Desulfovibrio orientis*	Fat curved rods, 5 μm × 1.5 μm. Hydrogenase apparently absent	
Dt. *ruminis*		Rods, 3–6 μm × 0.5 μm. Growth on pyruvate in sulphate-free media. Found in rumen of sheep	

in York.[39] These preserved objects were covered with a thin compact and strongly adherent black film of ferric and ferrous phosphates ($FePO_4.2H_2O$ and vivianite, $3FeO.P_2O_5.8H_2O$). Detailed accounts of the fundamental electrochemical processes are well documented in the literature[1,2] and the present discussion has been limited to just the extent required to interpret the microbiological involvement in the processes.

When a metal is immersed in water, damp soil and other aqueous systems the initial reaction is the dissolution of the metal as metallic cations which leave behind an excess of electrons:

$$Fe \rightarrow Fe^{2+} + 2e$$

These electrons are consumed at nearby cathodic sites by the balancing reaction, which in near-neutral solutions is usually the reduction of oxygen to hydroxyl ions

$$O_2 + 2H_2O + 4e \rightarrow 4OH^-$$

The overall reaction is the formation and subsequent precipitation of insoluble products formed by the reaction of ferrous ions with hydroxyl ion (and frequently oxygen) in solution. In the simplest case the reactions occur uniformly over the surface of the metal. The anodic sites are determined either by the heterogeneity of the surface or by random failure of a tenuous pre-existing oxide film. The overall rate of corrosion in these circumstances depends on a number of factors. The important criterion, however, is that the anodic and cathodic reactions must proceed in balance, to preserve overall electroneutrality. In the absence of oxygen the usual cathodic reaction for corrosion processes is the reduction of hydrogen,

$$2H^+ + 2e \rightarrow H_2$$

a reaction is rapidly limited in neutral solutions thus stifling the corrosion process. At lower pH values hydrogen evolution usually predominates both in the presence and absence of oxygen.

However, in addition to these basic reactions there are other factors which have a bearing on the overall process, in particular the microgeometry of the metal surface and variation in the oxygen concentration of the environment creating differential aeration cells. The cathodic sites in this case are the more oxygenated regions. Similarly if the metal is in contact with a soil or solution in which concentrations of aggressive anions vary from one site to another a concentration cell is produced.

These phenomena usually encourage localized pitting corrosion, which is also exacerbated by the action of certain microbiological agencies.

Originally, many workers in the field[4-7,15] considered that for micro-organisms to participate in the corrosion process they must, in general, be growing rather than dormant, and generating oxidizing agents by their metabolism. As an alternative, their role was to participate directly in one or both of the electrochemical reactions on the metal thereby initiating or exacerbating the reaction. Similarly it has always been accepted that microbial growths or colonies, whether growing or not, could be responsible for creating differential aeration cells on the surface of the metal. In some cases the microhabitats generated by these growths created anaerobic conditions which encouraged the activity of the sulphate-reducing bacteria. As a result, previous reviews[6,7,40] of the subject have tended to classify microbial corrosion on the basis of the metabolism of the organisms as follows:

1. Absorption of nutrients (oxygen) by microbial growths adhering to the metal surface.
2. Liberation of corrosive metabolites or end products of fermentative growth.
3. Production of sulphuric acid.
4. Interference in the cathodic process under oxygen-free conditions by obligate anaerobes.

Although this classification is useful, it seems to be more helpful to engineers to discuss the problem in terms of either aerobic or anaerobic environmental conditions, thus avoiding the more difficult biochemical aspects of the subject, and the mechanisms of microbial corrosion are discussed here under these two major headings.

5.2 Corrosion Under Aerobic Conditions

5.2.1 Colonization of Materials by Microbial Growths and Deposits

A wide range of aerobic organisms are able to fulfil this role. In general they do not produce aggressive metabolites but are responsible for the formation of differential aeration cells. In other words, they deplete a nutrient source thus producing an aggressive environment. This particular problem is often described as biofouling, that is the deposition on surfaces or blockage of flow channels and pipes by materials of biological origin. This may involve a variety of organisms, in particular, filamentous fungi, algae, protozoa and diatoms. It also includes the highly hydrated gelatinous slimes produced by the organisms. Microbial fouling of the surface almost invariably precedes such macrofouling. In the marine environment developed fouling usually involves barnacles, mussels, seaweed, etc. All types of microbes which colonize a surface must therefore be considered potentially dangerous. Typical examples of the microflora identified include the filamentous algae Chlorophyta and Cyanophyta bacteria of the species *Pseudomonas aeruginosa* as well as species of *Flavobacterium, Aerobacter, Gallionella* and *Sphaerotilus.* Fungal genera often identified include *Trichoderma, Monilia* and *Penicillium.*

The mechanism of the corrosion is simply the formation of a differential aeration cell due to the uptake of oxygen by the microbial colony, the oxygen concentration under such a mass becoming depleted. The poorly aerated surface becomes the anode of the cell while the better aerated regions, away from the deposits, provide the balancing cathodic reaction. This mechanism is illustrated in Fig. 2. A classic example of this type of problem is the fouling of industrial cooling water systems and heat exchangers. Contamination of the water by airborne aerobic bacteria and spores is common, particularly in the cooling towers of such plants. In addition, because these systems are operated on a continuous bleed and make-up basis, once contaminated the problem will continue to develop until positive remedial measures are taken. Another example of this type of problem is the external fouling of steel structures in industrialized river estuaries and other environments. The formation and retention of accretion in these circumstances is influenced by water movement and turbulence, and it is often difficult to assess the relative importance of microbial and other factors in the overall corrosion pattern.

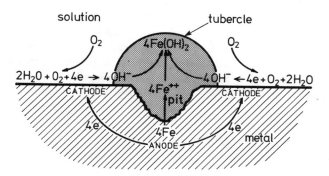

FIG. 2. Oxygen concentration cell pitting. After Iverson.[6]

5.2.2 Corrosion Due to the Formation of Tubercles

This is another form of the problem discussed above. It is mainly concerned with potable water distribution pipelines in which the formation of tubercles is encouraged by the presence of the iron-oxidizing bacteria *Gallionella, Sphaerotilus* and *Leptothrix*. These organisms obtain their energy for carbon dioxide fixation by the oxidation of ferrous ions to ferric ions, with the consequent accumulation of ferric hydroxide on the internal surfaces of the pipeline. This usually consists of a hard excrescence which shields the surface of the pipe from contact with the oxygen dissolved in the water. The relatively small area of metal beneath the tubercle becomes the anodic site, which in these circumstances sustains severe localized corrosion, that often results in perforation of the pipe wall. Again, tuberculation, like microbial slimes, provides a suitable habitat for the sulphate reducers, so that corrosion can be promoted by both mechanisms.

Recent studies by Ainsworth *et al.*[32] have confirmed that tubercles consist of a thin shell of hard magnetite (Fe_3O_4), beneath which is a mixture of hydrated ferrous oxides and iron sulphides. The surface of the tubercle in contact with the water is covered with a thin layer of goethite ($\alpha FeO(OH)$) in conjunction with compounds adsorbed by the tubercle from the water, i.e. organic material, silica, manganese. In many cases the magnetite layer contains a high concentration of carbonate, in excess of the calcium equivalent, and under these conditions the carbonate is believed to be present as siderite ($FeCO_3$). Within the tubercle the corrosion product may contain up to 5% of sulphur as sulphide. An interesting feature in this study was the fact that the population of sulphate reducers is higher on the surfaces of the tubercle than in the interior, and in some cases the interior can be free from these organisms. It has been suggested that this occurs because the sulphide inhibits the activity of the organisms, and sulphate reduction only takes place near the surface of the nodule, while the resulting sulphide diffuses back into the nodule to react with ferrous ions and be precipitated as iron sulphide. The significance of these findings will be discussed in more detail under anaerobic corrosion.

5.2.3 Corrosion Due to Acidic Environments

The corrosion problems experienced by industry under this heading fall into two categories, those due to the presence of sulphuric acid and those due to the presence of organic acids.

Corrosion which occurs in an environment containing a substantial amount of sulphuric acid can usually be attributed to the activity of autotrophic bacteria of the genus *Thiobacillus*. Detailed accounts of the chemistry of these organisms have already been mentioned. The significant feature of this type of corrosion is that it is an indication that there is a local supply of sulphur or one of its reduced compounds. The possible sources of these materials have been discussed above in Section 4.3.1.

The most important single problem caused by these organisms is the corrosion or degradation of concrete sewers. The mechanism is interpreted as the onset of septicity leading to the release of hydrogen sulphide gas, which is absorbed by the surface of the sewer where it can be oxidized by bacterial action to form sulphuric acid, which subsequently degrades the fabric of the sewer. A fuller treatment of the chemistry has been provided by Thistlewayte[41] and others.[5,30,42] This type of degradation has caused the collapse of concrete cooling towers, and there have also been a number of cases of corrosion of building stone and archaeological structures such as the Temples of Cambodia,[31] and the Parthenon in Greece. Failures of ferrous pipes by similar mechanisms have also been widely reported.[5] The situation in which acid production by *Thiobacillus* spp. follows the production of sulphur by sulphate reducers in an adjacent environment has already been mentioned, but perhaps less expected was a recent very serious problem which arose when an apparently innocuous backfill was infected by both types of organism as a result of ground water movement. However, not all of the activities of *Thiobacillus* spp. are detrimental; the iron-oxidizing bacteria of the species *Thiobacillus ferrooxidans* are used successfully for the recovery of metals from low-grade ores by leaching routes, although the acidic environment created by their activity may cause severe corrosion problems for mining machinery and other equipment. Detailed accounts of the bacterial leaching of ores have been published by Le Roux[43] and Fletcher.[44]

When corrosion can be attributed to the formation of organic acids it is nearly always attributable to the presence of fungal growths. A number of species of fungi are capable of excreting organic acids during their normal oxidative metabolism, one of the more important of which is *Cladosporium resinae*. This organism is predominantly responsible for the corrosion which occurs in the aluminium integral fuel tanks of aircraft. The primary factor considered to be responsible for this problem is condensed water in the kerosene. The fungal spores present in the water germinate at the water/kerosene interface by oxidative assimilation of the kerosene, inorganic constituents dissolved in the water providing the nutrient source for growth. During growth the fungus liberates organic acids into the water phase producing sufficient acidity to encourage localized corrosion. The ability of the fungus to utilize carbon from the kerosene for microbial metabolism is considered to be a primary factor in fuel microbiology, and experience indicates that the C_3–C_{16} alkanes of aviation fuels are particularly susceptible to degradation. Sulphate-reducing bacteria have also been reported as a possible causative agent for this type of corrosion of aluminium,[45-47] and the formation of differential aeration cells by the fungal growth on the surface of the metal may make a substantial contribution in some cases. A similar problem is experienced in metal machining and forming where the oxidation of constituents present in oil emulsion lubricants may cause severe corrosion of tools, etc.[24]

Many of the organic acids produced by fungi have been demonstrated to be corrosive to metals by Copenhagen,[48] and Calderon et al.[49] The latter authors consider that for metals which can evolve hydrogen during the corrosion process, the organic acid is reduced according to the equation

$$2\,[\mathrm{H^+_{acid}org^-}] + 2e \rightarrow \mathrm{H_2} + 2\mathrm{org^-}$$

which is followed by precipitation of an organic salt of the metal

$$\mathrm{M^{2+}} + 2\mathrm{org^-} \rightarrow \underset{\text{salt}}{[\mathrm{M(org)_2}]}$$

For metals with a lower oxidation potential than hydrogen, the same product is achieved by neutralization of the organic acid by cathodically produced alkali:

$$\mathrm{H_2O} + \tfrac{1}{2}\mathrm{O_2} + 2e \rightarrow 2\mathrm{OH^-}$$

$$2[\mathrm{H^+_{acid}org^-}] + 2\mathrm{OH^-} \rightarrow 2\mathrm{H_2O} + 2\mathrm{org^-}$$

$$\mathrm{M^{2+}} + 2\mathrm{org^-} \rightarrow \underset{\text{salt}}{[\mathrm{M(org)_2}]}$$

5.3 Corrosion Under Anaerobic Conditions

5.3.1 An Overview
From a consideration of the electrochemical reactions discussed earlier, significant corrosion of ferrous metals would not be expected in an oxygen-deficient near-neutral environment, since neither of the two common cathodic reactions can occur at a substantial rate. However this

is not always the case, and when it does occur the severity of the corrosion can be orders of magnitude greater than that experienced under normally aerated conditions. The primary factor responsible for this situation is the activity of the sulphate-reducing bacteria and their sulphide metabolic products, and the corrosion of buried metals by these mechanisms is of great economic importance.

The mechanism postulated by von Wolzogen Kühr and van der Vlugt[15] for the involvement of bacteria in the corrosion process, suggested that cathodic hydrogen is utilized for the dissimilatory reduction of sulphate. The equations to describe this suggestion are given below:

$$4\ Fe \rightarrow 4\ Fe^{2+} + 8e \qquad \text{—anodic reaction}$$
$$8\ H_2O \rightarrow 8\ H^+ + 8\ OH^- \qquad \text{—dissociation of water}$$
$$8\ H^+ + 8e \rightarrow 8\ H \qquad \text{—cathodic reaction}$$
$$SO_4^{2-} + 8\ H \rightarrow S^{2-} + 4\ H_2O \qquad \text{—cathodic depolarization by bacteria}$$
$$Fe^{2+} + S^{2-} \rightarrow FeS \qquad \text{—corrosion product}$$
$$3\ Fe^{2+} + 6\ OH^- + 3\ Fe(OH)_2 \qquad \text{—corrosion product}$$

$$4\ Fe + SO_4^{2-} + 4\ H_2O \rightarrow 3\ Fe(OH)_2 + FeS + 2\ OH^- \text{—overall reaction}$$

If the mechanism were correct, the ratio of corroded iron to iron sulphide should be 4:1. This is seldom found in practice and wide deviations occur ranging from 0.9:1 to nearly 50:1. Considerable research has been performed over the years to support this mechanism,[50-53] but recent studies indicate that an alternative mechanism involving the metabolite sulphide and the presence of iron sulphide films may be more important.[54-56] It is interesting to record that laboratory studies using growing cultures of sulphate-reducing bacteria seldom produce the severe pitting corrosion experienced in the field, and that the normal corrosion products of iron oxide or hydroxide are nearly always absent. The same also applies to the presence of elemental sulphur which often forms under natural conditions around the periphery of pitted zones, but which is seldom detected in the laboratory.

The corrosive nature of solid and dissolved sulphides is well known; it has been reviewed by Smith and Miller.[57] Several authors[58-63] have reported high rates of corrosion, typically rapid perforation of water distribution systems, due not only to dissolved sulphides but also to iron sulphide and indirectly to sulphate-reducing bacteria. Again, the significance of elemental sulphur as a corrosive agent as an alternative to the action of the bacteria has recently been discussed by Schaschl.[64] An alternative theory is supported by the fact that selective biochemical examinations[66] using different growth and nutritional media with an alternative to sulphate as electron acceptor have not reproduced some of the earlier results, i.e. little if any cathode depolarization occurred. In addition, weight-loss tests have generally not been convincing in support of the classical theory.[54,65,66] The combination of these observations and others reported more recently by King and Miller[55] has led to a more critical approach to the classical theory. The situation is clearly more complex than had originally been envisaged and it is helpful to discuss both mechanisms separately at greater length.

5.3.2 Classical Theory

Support for the classical theory has been based on the understanding that certain strains of the sulphate-reducing bacteria possess the enzyme hydrogenase.[66] This enzyme is able to catalyse both the reversible reactions involving hydrogen

$$H_2 \rightleftharpoons 2\ H \rightleftharpoons 2\ H^+ + 2e$$

and the reduction of sulphide by molecular hydrogen according to the equation

$$SO_4^{2-} + 4\ H_2 \rightarrow S^{2-} + 4\ H_2O$$

The amount of hydrogenase possessed by an organism can be determined quantitatively by manometric biochemical techniques and is usually reported in terms of the hydrogen absorption coefficient, $-Q_{H_2}$ (microlitres H_2 absorbed per milligram dry weight of cells per hour), the significance of which will become apparent.

Booth and Tiller,[67-69] using a simple galvanostatic polarization technique, showed that growing strains of hydrogenase-positive mesophilic, halophilic, and thermophilic bacteria had the capacity to depolarize a cathode. Hydrogenase-negative organisms were completely inactive in this process. Similar results were obtained by Horvath and Solti,[70] but it was Iverson[71] who produced more positive evidence in support of the mechanism originally suggested. By placing a pair of steel electrodes in an agar plate containing an organic buffer and a redox dye (benzyl viologen) as the electron acceptor, he demonstrated that a hydrogenase-positive strain of the bacteria could utilize cathodic hydrogen for the reduction of benzyl viologen from colourless to a purple colour. This was achieved by placing a heavy concentration of the organism on an area of the agar in contact with one electrode and incubating for a period of 9 h under nitrogen. When the electrode was removed the dye had been reduced, while a concentration of ferrous ions had developed under the electrode which was not in contact with the organism.

Subsequent studies by Booth et al.[72] using a potentiostatic technique and washed, resting (non-growing) cells in an inert organic buffer with benzyl viologen as the electron acceptor, confirmed the previous results. Typical results are illustrated in Fig. 3. The procedure involved monitoring the rate of change in current density due to the reduction of the dye.

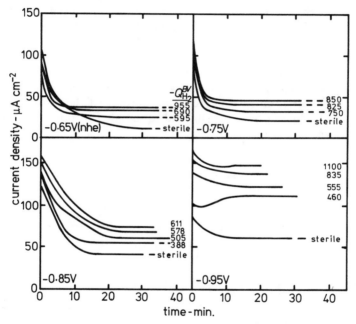

FIG. 3. Current density/time curves for *Desulfovibrio vulgaris* system with benzyl viologen as electron acceptor. After Booth and Tiller[72], Crown Copyright.

In the absence of a reducible substrate the polarization of the cathodes was independent of the hydrogenase activity of the bacteria whereas when such a material was present the cathodic depolarization was increased by the hydrogenase activity. The rate of reduction of the substrate was a function of the hydrogenase activity and a logarithmic relationship existed between the additional current to maintain a cathode at various potential values in the presence of the bacteria above that required in buffer solution alone (Δi) and the enzymic activity of the organism (Fig. 4). By plotting the slopes of these lines (d log Δi/d $Q_{H_2}^{BV}$) against the controlled potential a straight line is obtained which intercepts the potential axis at -0.98 V NHE indicating that the rate of depolarization of the cathode becomes constant irrespective of the enzymic activity (Fig. 5) of the bacteria. These results supported the theory that sulphate-reducing bacteria act as depolarization agents. However this experimental procedure has been criticized because recent studies[7,41] have shown that these organisms often retain iron sulphide attached to their cell walls. Since the cultures had been grown in a standard medium the resting cell suspensions were probably already contaminated with iron sulphide, which could assist in the depolarization process. However, disrupted cells and cell fractions were shown to be completely inactive in depolarizing the cathode,[72] which is somewhat surprising if iron sulphide is the active constituent. On the other hand, experiments carried out with cytoplasm reproduced the results of those using resting (non-growing) cells, indicating that the location of the active enzymes is in the cell cytoplasm.

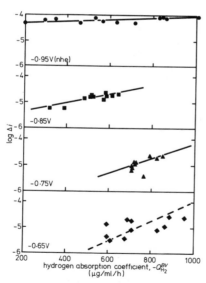

FIG. 4. Relationship between 'bacterial current' and hydrogenase activity for *Desulfovibrio vulgaris*. After Booth and Tiller[72], Crown Copyright.

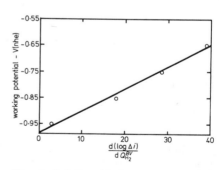

FIG. 5. Relationship between potential and $\dfrac{d(\log \Delta i)}{dQ_{H_2}^{BV}}$ for *Desulfovibrio vulgaris*. After Booth and Tiller[72], Crown Copyright.

From the experimental evidence collected by various workers it was concluded that for cathode depolarization to occur the bacterium must contain the enzyme hydrogenase. Strains such as *Desulfovibrio desulfuricans* and *Desulfovibrio vulgaris* contain the enzyme and so utilize the cathode hydrogen for the reduction of a suitable substrate if required, although this is not obligatory. Hydrogenase-negative strains such as *Desulfotomaculum orientis* have been shown to be completely inactive in this mechanism. It was also suggested that there was a direct relationship between the hydrogenase coefficient of the organism and the magnitude of the depolarization, and hence possibly the rate of corrosion. The rate of corrosion of steel is thus demonstrably related to hydrogenase activity but only under a limited range of experimental conditions.[55] Subsequent experiments using actively growing cultures of a number of different strains of sulphate-reducing bacteria indicated that the rate of corrosion was independent of hydrogenase activity.[73,74] In addition, the hydrogenase-negative organism showed itself to be quite as aggressive as the *Desulfovibrio* species. These data clearly indicated that other factors must also be involved, as will appear.

5.3.3 Alternative Cathodic Depolarization Mechanisms

Reference has already been made to the fact that both dissolved and solid sulphides as well as elemental sulphur itself are recognized as corrosive agents, and that iron sulphide is cathodic to mild steel. A quantitative study of the corrosivity of iron sulphide was carried out by Stumper,[75] who reported that the rate of corrosion of mild steel increased by approximately 230% by contact with iron sulphide and was doubled when the mild steel was not in direct contact with the iron sulphide but was connected to it by a wire. Similarly it was reported by Sheppard[61] and Treseder[76] that the serious corrosion which occurs in sour gas environments could be attributed to the presence of sulphide compounds and elemental sulphur, and commented that the rate of corrosion varied depending on the species from as little as 5.3 mm annually for iron sulphide to 1100 mm annually for sulphur. Elemental sulphur has also been demonstrated to be very aggressive by Farrer and Wormwell.[77] More recently Schaschl[64] has discussed the corrosive nature of elemental sulphur in deaerated neutral solutions, and considers it to be more important than the role of the sulphate-reducing bacteria.

Laboratory studies[78] of the influence of ferrous iron on the anaerobic corrosion of mild steel by actively growing cultures of sulphate-reducing bacteria suggested that protective films were formed on the metal surface at low ferrous iron concentrations, while at concentrations sufficiently high to precipitate all the bacterially produced sulphide as iron sulphide no protective films were formed and the rate of corrosion increased markedly.[79] Electrochemical studies showed that sustained and vigorous cathodic depolarization occurred even if the ferrous iron concentration was subsequently reduced. In an earlier study[80] with bacteria growing in a medium containing 0.5% ferrous iron, the corrosion rate of mild steel was approximately a linear function of time and independent of the hydrogenase activity of the bacteria. The results of complementary weight loss experiments and polarization studies clearly indicated that the ferrous sulphide mechanism made a major contribution to the overall corrosion. Comparisons of the corrosion of mild steel by chemically prepared iron sulphides and biogenic iron sulphide indicated that in media with high ferrous iron concentrations most of the corrosion in the bacterial culture can be attributed to the biogenic ferrous sulphide.

King et al.[81] showed that the corrosivity of chemically prepared iron sulphide is a function of the sulphur stoicheiometry, but that the stimulation of corrosion by iron sulphides decreases with time. High rates of corrosion in bacteria-free systems can only be obtained by continual replenishment of the iron sulphide.

Cathodic polarization studies using chemically prepared and biogenically prepared iron sulphides have confirmed that these materials are capable of utilizing cathodic hydrogen.[54] It has also been shown that there is a linear relationship between the amount of chemically prepared iron sulphide added to a bacteria-free system and the additional current required to maintain a cathode at a particular controlled potential,[82] a result in agreement with those by Booth and Tiller[72] for resting (non-growing) cells. The decrease in the efficiency of depolarization by iron sulphide with time in the absence of bacteria is attributed to the binding of atomic hydrogen within the iron sulphide lattice. However the activity of the iron sulphide can be restored in the presence of sulphate-reducing bacteria because of their ability to utilize or disbond the hydrogen contained in the film. Thus an explanation is provided for an alternative mechanism which is considered to be responsible for the continued high rates of corrosion observed in soil environments. A schematic diagram illustrating the two mechanisms is given in Fig. 6.

One of the important features in this discussion is the nature of the sulphide film. Reference has already been made to the role of ferrous iron in the formation of protective or non-protective films. The chemistry of the formation of six stable iron sulphide minerals has recently been

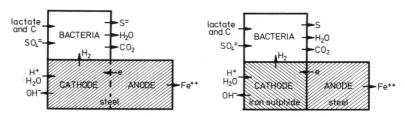

FIG. 6. Schematic representation of the classical mechanism (left) and alternative mechanism (right) of microbial corrosion under anaerobic conditions. After Miller and King.[7]

studied in considerable detail by Rickard[83] and also by Mara and Williams.[56] These latter studies confirmed the earlier views of King et al,[81] but emphasized that the severity of microbial corrosion can be explained by the nature of the sulphide films formed on the surface. The primary corrosion products appear to be mackinawite and siderite. Only the latter is protective, but both are transformed by sulphidation under anaerobic conditions to greigite and smythite, with eventual transformation to pyrrhotite which is particularly non-protective. The chemistry of these processes is illustrated in Fig. 7. Under aerated conditions elemental sulphur may be formed. Recent reports[64] indicate that in deaerated solutions this element acts in the same manner as dissolved

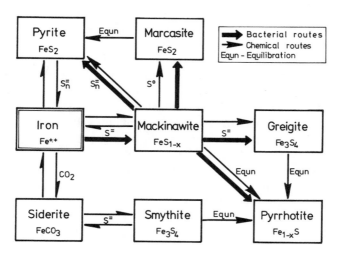

FIG. 7. Interrelationships between the sulphides of iron. After Rickard.[83]

oxygen in promoting the corrosion of steel. Sulphur is solubilized by sulphide irons, and the solubility is also affected by pH and temperature. The sulphur appears to promote corrosion by a concentration cell mechanism similar to that described for differential aeration cells, i.e. an anodic area can develop underneath a porous material which shields the steel from dissolved sulphur, the corresponding cathodic area being the adjacent region where dissolved sulphur is readily available as a cathodic reactant. In this situation it is suggested that the bacteria may merely provide the shielding action needed to promote a concentration cell effect.

The alternative mechanism outlined above provides an attractive interpretation for much of the phenomenology of microbiological corrosion under anaerobic conditions, and it has gained considerable support. However it must be emphasized that cathodic depolarization of a mild steel surface by this means only occurs when the metal is in the horizontal plane and the sulphide is resting on the surface. There is no evidence to suggest that the mechanism will be effective under conditions similar to those described by Booth and Tiller[72] for their resting cell studies. It is also interesting to note that Munger[84] reports that in the practical situation severe sulphide pitting in oil tankers carrying sour crude only occurs in the lateral (horizontal) members of the bulkheads.

6. ASSESSMENT OF CORROSIVITY AND PREVENTION AND CONTROL OF MICROBIAL CORROSION

6.1 Diagnosis of Corrosivity

Because of the complexity of many microbial problems once established, it is important to give consideration to the prediction of situations in which problems may arise, with a view to avoiding them wherever possible. The range of environments which must be considered is wide, and the criteria for prediction are not equally well established for all situations. Nevertheless a reasonable approach can be made to prediction in many practical cases, and neither of the alternatives of installing a complex protection scheme when not required, or ignoring possible risks until problems arise is economically acceptable when accepted predictive procedures are available.

6.1.1 Corrosivity of Soils and Sediments

Although the microbiological condition of the soil is important, other factors also need to be assessed. According to a number of workers[85-87] the main factors responsible for making a soil aggressive are anaerobic bacteria, the soil chemistry (particularly the acidity and salt content), the redox potential of the soil, soil resistivity, differential aeration and the formation of concentration cells. These are not all independent, and different workers attribute weight to various combinations of factors. Starkey and Wright[86] consider that redox potential is important since it measures the risk of microbiological activity by sulphate-reducing bacteria and they advocated guidelines based on those given in Table 2.

TABLE 2

RELATIONSHIP BETWEEN SOIL REDOX POTENTIAL AND CORROSIVENESS (STARKEY AND WRIGHT[86])

Soil E_H redox potential H_2 scale (mV)	Assessment of corrosivity
100	Severe
100–200	Moderate
200–400	Slight
400	Non-corrosive

However others[88] prefer to use soil resistivity data since they have been shown to correlate well with the corrosivity of soils. Complementing these two criteria, evidence has been produced to show the importance of pH and water content,[89] and there has also been an effort to correlate counts of viable sulphate-reducing bacteria with corrosivity. Studies at Teddington carried out by Booth et al.[89,90] have clarified the situation, and a simple testing scheme devised has found favour with industry. The scheme involves the determination of three parameters: (i) soil resistivity; (ii) redox potential; and (iii) water content. The potential corrosivity of the soil can than be assessed by applying the scheme of Table 3.

During this study the importance of soluble iron and hydrogen uptake of the soil was also examined. The latter parameter varied considerably and was found to be unreliable. However the significance of soluble iron in the soil was established and soils with mean soluble iron content of ≥ 120 mg/g were found to be aggressive. This estimation is time consuming, and its use in diagnosis is not recommended, however the concentration of soluble iron is important in governing the nature of any

TABLE 3

SCHEME FOR ASSESSMENT OF RESISTIVITY OF SOILS (BOOTH AND TILLER[72])

	Aggressive	Non-aggressive
Soil resistivity (Ω cm) and/or	$<2\,000\,\Omega$	$>2\,000\,\Omega$
Redox potential at pH 7 (V(nhe))	<0.40 (or <0.43 if clay)	>0.40 (or >0.43 if clay)
Borderline cases to be resolved by water content (wt%)	$>20\%$	$<20\%$

sulphide film formed on the metal surface which may modify the overall rate of corrosion.

The reliability of these predictive tests was confirmed by a major burial programme carried out on mild steel, aluminium, copper, zinc and lead.[91] The relationship between soil type and corrosion was clearly identified. According to a recent survey[11] soil resistivity is the most favoured single test used. The usefulness of the redox potential determination is accepted but it is not used as frequently as it might be because of the fragile nature of the instrument and a problem with reliability in unspecialized hands. A new design of probe which overcomes these problems will shortly be commercially available.

To complement this type of test procedure a wide range of simple microbial assay methods is available. These are particularly valuable for monitoring the biodeterioration of cutting oils and emulsions, for example, but they are not at present accepted for assessing soils and waters. They are in general used to corroborate the other tests or when ambiguous test data are obtained. Several media and techniques have been described by Miller and Tiller[37] while various standard media have been evaluated by Mara and Williams.[92] The disadvantage of these tests is that they will frequently indicate the *presence* of sulphate-reducing bacteria in soil and water samples irrespective of whether or not the environment is likely to support their *activity*. There are many situations in which the presence of the bacteria is known but because the environmental conditions are unsuitable for proliferation and activity aggressive conditions are not likely to arise unless significant changes occur. Such changes in environmental conditions can occur over the years and a soil classified as non-aggressive by any test method may develop into an aggressive site at a later date. One of the factors which should never be overlooked is that the trenches in which pipelines have been laid may often act as land drains, and be liable to partial flooding which may encourage the development of anaerobic conditions.

With the increase of offshore activity, consideration of seabed sediments has increased in significance, and King[93] has recently devised a procedure to assess their corrosivity. In this case the carbon, nitrogen and phosphorus contents of the sediment are the significant criteria, in combination with soluble iron contents. By adopting simple assessment procedures it has been possible to prepare corrosivity maps of various sections of the North Sea thus providing valuable guidance for the design engineer.

Certain problems remain to be resolved. In particular the assessment of the aggressiveness of undisturbed soil at depths in excess of 3–5 m poses difficulties. In nearly all cases severe microbial corrosion occurs only in disturbed soils and much of the available data refer to structures buried at depths between 2 and 3 m. According to Morley[94] however, undisturbed soils do not constitute a problem and data obtained from driven steel piles which have been pulled from a variety of environments have not been found to indicate sustained severe localized corrosion. This view is shared by workers in the USA[95] and Australia,[96] although the Japanese[97] and Scandinavians[98] have reported severe pitting and perforation of steel piles at depths in excess of 4 m. The problem which confronts engineers is therefore to what extent is it necessary to determine the corrosivity of soils at such depths, and how best to utilize existing testing procedures to do so. With experience it is possible to adopt the testing procedures outlined above to borehole samples and by utilizing published data to provide acceptable corrosion information for design needs. From a microbiological point of view the argument for inactivity of undisturbed soil has been based entirely on the nutritional requirements of the organisms. However Postgate[36] comments that the organisms have been isolated from undisturbed ancient geological strata where they have lain dormant, but still viable given favourable environmental changes. To what extent the disturbance of the soil by a driven steel pile is sufficient to produce such changes is one of the remaining points of discussion in this area.

6.1.2 Corrosivity of Waters
There are well-established criteria for assessing the general corrosivity of waters, which are not relevant here. It has already been shown that many species of bacteria can be isolated from most natural waters whether

fresh or salt. The problem of predicting whether microbial factors will make a significant contribution to the rate or distribution of corrosion depends on the assessment of the probability of conditions developing in which microbial activity can proliferate. In marine conditions, sulphate-reducing bacteria can often thrive in the anaerobic microclimate which can develop beneath corrosion tubercles and marine fouling, and produce a significant increase in corrosion rate and pitting propensity. Problems arise less readily in fresh water situations, although the activity of sulphate-reducing bacteria can often be demonstrated beneath corrosion tubercles in iron water mains, where because the rate of general attack may be very low the contribution to pitting and graphitic corrosion by the bacterial activity may be relatively more significant.

Estuarine conditions pose certain predictive problems but simple procedures can be used to approach the assessment of potential risk. Again, the sulphate reducers pose the most serious problems. Studies in the Thames[99-101] and the Afon Dyfi[102] estuaries have shown that corrosion under these conditions is influenced by variations in salinity and oxygen concentration and flow velocity. In nearly all cases, and in particular in the highly aerated Afon Dyfi, specimens developed encrustations and tuberculation. Beneath these growths extensive pitting corrosion of the metal surface occurred and an analysis of the corrosion products showed that in nearly all cases it comprised mixed oxides and sulphides of iron and in some cases iron sulphate. In some instances these coatings conferred protection to the underlying metal. The corrosion product on metal plates buried in the mud was primarily a mixture of FeS, $FeSO_4$ and Fe_3O_4. Complex situations therefore exist in these types of environments, but by monitoring the concentration of oxygen and other inorganic species such as sulphide and variation in salinity, it is possible to acquire data which clearly indicate the possibility for microbial corrosion to occur even in waters which are aerated in the bulk.

6.2 Prevention of Microbial Corrosion

There is no universal approach to the prevention of corrosion by microorganisms and the problem is usually tackled more from the corrosion than the microbiological point of view. There are a number of methods which can be adopted to prevent this type of problem, that may be discussed typically for a pipeline situation, but which may be modified where appropriate for most buried metal problems.

6.2.1 Provision of a Non-Aggressive Backfill

Protection may be afforded by surrounding a pipe with a non-aggressive backfill such as chalk or sand, provided that the trench is well drained. Chalk primarily provides an alkaline environment, producing a sufficient pH change to inhibit any activity or growth of the sulphate reducers. A range of sands is available, and these are useful in as much as they provide a physical barrier capable of uniform compaction. However, they provide little protection against subsequent infection by water-borne organisms or contamination by metabolic products arising from activity in adjacent regions. The addition of sparingly soluble biocides to the backfill has been suggested in some instances by Purkiss and Cameron,[103] but this can be an expensive form of treatment. Proprietary inert backfilling materials are also available, though usually at greater cost than natural materials.

6.2.2 The Use of Protective Coatings and Tapes

These systems provide a barrier between the metal and its environment impervious to soluble aggressive species. The concept of using protective coatings is very old, yet even with the many advances which have been made in this field it is doubtful if an ideal coating is available. The range of materials available is large and includes such products as coal tar enamels, asphaltic bitumens, epoxy resins and coal tar/epoxies, thermoplastic polymers, and cementitious materials. Detailed discussion of the various systems is outside the scope of this review, and the reader is advised to consult the specialized publications which are available, in particular the material which has been presented at a number of recent conferences.[104-106] Valuable information on the design and selection of specific coating systems can be obtained by reference to BS 5493:1977. The manufacturers of protective coatings and tapes also provide detailed information on their own products, and a booklet dealing with the suitability of coatings for specific environments is available from the British Steel Corporation.[107]

Most modern protective coatings will provide a very high standard of corrosion protection unless selected for applications in which they are specifically not suited. Nevertheless failures do occur and in many instances these can be attributed either to inadequate preparation or poor application. In some instances plastic tapes and impregnated wrappings have failed because they have been poorly applied by inexperienced operatives; inadequate overlap is a common fault. The most frequent cause of failure in these systems, however, is mechanical damage during delivery and laying or resulting from lack of care and supervision during backfilling.

In selecting any of these systems as a protection against microbial corrosion, it is advisable to confirm that the material itself is not susceptible to biodeterioration, as can occur on certain bitumastic coatings and tapes and other cellulosic or plastics materials.[108]

A recent development in the use of plastics has been the advent of loose polythene sleeving for pipelines. It is rather early to comment upon how effective this system will be in the UK, although the technique has been in use for many years in the USA, where it is considered to be performing well. The main doubts regarding the use of the method in the UK concern the possible effects of ingress of contaminated water at regions of mechanical damage. Nevertheless a number of water and gas authorities are adopting this procedure for their distribution schemes.

6.2.3 The Application of Cathodic Protection

The principles and theory of cathodic protection have been extensively discussed in the literature[2,109] and it is not intended to comment on them further here. Two techniques are commonly used to provide cathodic protection, the use of sacrificial anodes of magnesium, zinc or aluminium alloys which corrode preferentially and provide protection for the structure, and the use of impressed current from a rectifier, via an inert anode. The choice of method depends on the site conditions and the demands of the system.

Experience has shown that in the absence of sulphate-reducing bacteria protection of steel is achieved when the potential is depressed to -0.85 V (v. $Cu/CuSO_4$). When a soil environment is classified as aggressive and there is a high risk of microbial corrosion, the potential of the structure must be depressed by a further 0.10 V, i.e. to -0.95 V ($Cu/CuSO_4$) to achieve the desired protection. This additional requirement has been confirmed in practice and by the studies of Booth and Tiller[72] and of Horvath and Novak.[110] Depressing the cathode potential to these values has the effect of increasing the alkalinity at the metal/environment interface, which suppresses the growth of the sulphate-reducing bacteria. Caution must be exercised, however, since if the potential is depressed to still more negative values, alkaline blistering of protective coatings can occur.

Electrochemical studies[72] on the cathodic characteristics of mild steel in the presence of resting cell suspensions of sulphate reducers identified a linear relationship between the hydrogenase enzyme activity of the organism and the additional current required to maintain various cathodic potentials, illustrated in Figs. 4 and 5. A significant feature of this study was the accelerating effect of bacteria on the corrosion of an incompletely protected specimen (Table 4). A linear relationship was

TABLE 4

CATHOLYTE IRON ANALYSES AND CORROSION RATES OF 'CATHODE' AT VARIOUS CONTROLLED POTENTIALS (BOOTH AND TILLER[72])

Controlled potential of cathode	Catholyte iron (μg/h)		Corrosion rate of cathode (mg/dm^2/day)	
V(nhe)	Buffer	Buffer + Dv. vulgaris	Buffer	Buffer + Dv. vulgaris
−0.95	0	0	0	0
−0.85	25.5	33	3.1	3.9
−0.75	52.0	82	6.3	9.8
−0.65	112.5	221	13.5	26.5

established between potential of an incompletely protected cathode and the rate of corrosion occurring (Fig. 8). It can seen that to reduce the rate of corrosion in the presence of microbial activity to the same level as that observed on protected specimens in the absence of activity, an additional polarization of 0.1 V is required, as indicated above.

6.2.4 The Use of Non-Metallic Materials

The problem of underground corrosion could be avoided altogether by the use of chemically inert materials of adequate strength. A number of resistant alloys fall into this category, although the cost is uneconomic in most situations. Non-metallic materials are now gaining ground, particularly composite fibre-reinforced resin systems. Many of the initial problems associated with strength and durability of these products now appear to have been overcome and reliable systems are available for use in a wide spectrum of environments. Other plastic materials such as PVC and polyethylene have also been used. Care should be taken in selecting these latter materials since in some cases they contain plasticizers which are themselves susceptible to biodeterioration, resulting in progressive embrittlement of the polymer.

Cement products are widely used, including spun concrete and asbestos cement pipes and reinforced or prestressed concrete for buried

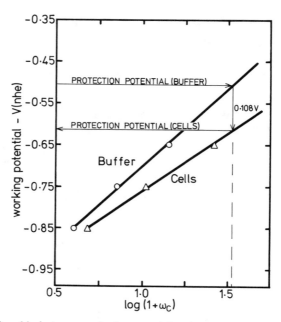

FIG. 8. Relationship between cathode potential and corrosion rate, illustrating the more severe cathodic protection criterion in the presence of sulphate-reducing bacteria. After Booth and Tiller[72], Crown Copyright.

piling, although these may have their own problems in some heavily contaminated soils and waters, making proper material specification and detailing important. Systems are also available in which fine concrete is sprayed on to both internal and external surfaces of a thin tubular former of Armco iron. Very large diameter pipes can be fabricated by this technique, which has had some success in aggressive soils in Middle East desalination plants.

6.3 Control of Microbial Corrosion

In some situations microbial infections will develop and cause problems in systems which were initially not aggressive, and the economic remedial measure may be to control the problem by the use of biostats or biocides rather than to adopt any of the methods outlined above. Again, in certain areas, infection and reinfection is a perennial and predictable problem and provision for chemical control should be included in the operating procedures from the outset. The use of biostats or biocides is discussed in relation to four specific situations. It is not intended here to discuss the mechanism by which these materials inhibit microbial activity; some compounds act as enzyme poisons (e.g. bisthiocyanate), while others are able to disrupt the bacterial cell wall and cytoplasm (e.g. cetyl pyridinium chloride and sodium dodecyl sulphate). The mechanisms and the resistance of various organisms to them have been reported by McCoy.[112] A detailed account relating to the sulphate-reducing bacteria has been published by Saleh et al.[113]

6.3.1 Aircraft Fuel Tanks

The primary problem in aircraft fuel tank corrosion is the presence of water, and it follows that every effort should be made to ensure that the ingress of water is reduced to a minimum and complete drainage is possible to remove that which inevitably appears. In addition, current practices involve the use of protective coatings such as nitrile rubbers, polysulphide rubbers and polyurethane. Unfortunately, fungi are capable of penetrating these coatings and other preventive measures must be considered. In some cases the inclusion of biocides in the tank coatings has been considered. To achieve the desired effects the material has to be compatible with the coating, only slowly leachable, and insoluble in the fuel. At present no biocide meets all these requirements adequately, and so the materials are normally dosed into the tank contents. Application is usually carried out using cartridges of water-soluble, fuel-insoluble biocides. Strontium chromate is popular in this application. The alternative is to use a hydrocarbon-soluble biocide which will build up to a toxic level in the water phase, typically ethylene glycol monomethyl ether (methyl cellusolve) at between 0.1% and 0.15% concentration. An alternative material is a proprietary organoborinane mixture used at a concentration of between 150 and 200 ppm. There are many unanswered problems in this field, and it is an area of active interest in the industry.

6.3.2 Cutting Oils and Emulsions

Ideally, good housekeeping and maintenance should normally avoid infection in these systems, but this is not always achieved and the use of biocides is common. In a straight oil the biocide must be sufficiently soluble to dissolve in the hydrocarbon phase and possess sufficient water solubility to migrate into water droplets as these occur. Chlorinated phenols are particularly useful in this instance although laboratory trials may be required to select the one most effective in a particular problem. In cutting oils and emulsion systems 'Panacide' has been found to be effective and compatible both in sterilizing plant and dosing emulsions. It is preferable to use continuous dosing systems rather than shock doses both to avoid detriment to the emulsion and to simplify the disposal problem.

6.2.3 Industrial Water Systems

Industrial cooling systems fall into three main categories, open re-circulatory systems utilizing cooling towers, closed systems and once through systems. All are susceptible to some type of microbial corrosion problems arising from the accumulation of bacteria, fungi, and algae. These are often accompanied by highly hydrated, glue-like gelatinous polymers produced by the organisms. In seawater cooling systems the situation may be more complicated by the growth of macrofouling, barnacles, mussels, etc, although microbial fouling generally precedes macrofouling.

The typical microflora of an open recirculatory cooling system will contain *Bacillus subtilus, B. cereus,* species of *Flavobacterium, Aerobacter, Gallionella, Sphaerotilus,* and the sulphate-reducing bacteria. Algae of the genus *Chroococcus oscillatoria* are also often present. All promote the formation of differential aeration cells, which encourage localized corrosion beneath the fouling.

Treatment of these systems, apart from the selection of alloys which are resistant to this type of phenomenon, such as the copper and nickel alloys, is by the use of biocides. Obviously certain practical factors must be considered. For example it is neither necessary nor possible to render a system completely sterile. In order to keep microbial growth to a minimum or an acceptable value the biocide must meet specific requirements. It must be economic at an effective concentration and safe to handle. It should be non-reactive to dissolved organic substances and dead microflora and must be compatible with the materials of fabrication of the plant. At the same time it must have a broad spectrum of activity and be stable in dilute solutions over long periods.

The range of commercial biocides is extensive, covering a wide range of chemical types, including the chlorinated phenols, quaternary ammonium compounds (relatively ineffective against fungi), gluteraldehyde, polyamines, acridine dyes, bisthiocyanate, bis-*p*-chlorophenyl diguanido-hexane diacetate, zinc dimethyldithiocarbonate, and 2,2-hydroxy-5-dichlorodiphenylmethane. Several reviews are available[111,112,113] on the use of the organic biocides and the concentrations required for successful control of biofouling and microbial corrosion. It is important to emphasize that these materials in no way dispense with the necessity to use conventional corrosion inhibitors and scale control chemicals. In many cases, a number of biocides may be selected, used in rotation to avoid the development of resistant strains. Comprehensive water treatment programming is offered by many commercial organizations, and specifications can be offered to overcome the majority of cooling water problems.

Chlorination is also a popular approach to sterilization in water systems; the concentration requirements vary but are usually of the order of 1 ppm of free chlorine. Chlorine must be used with caution, however, because in low concentrations (0.5–2 ppm) it can react with materials of construction. In addition, its volatility makes necessary either continuous dosing or frequent intermittent dosing. Chromate has been another popular inorganic biocide over the years, but its use is now being limited due to its toxicity and potential problems of availability.

6.3.4 Potable Water Transmission Pipelines

Control of the activity of the iron bacteria in potable water systems can be achieved acceptably by the use of chloramine or, of course, chlorine. This is of limited benefit in controlling the growth of tubercles, but it cannot remove the existing tubercles which lead to a reduction in flow capacity that is in many cases more serious than the corrosion itself.

7. SUMMARY AND CONCLUSIONS

The review has clearly indicated that there is a substantial body of information available concerning microbial corrosion, which in many instances is associated with the more common problem of biofouling. Obviously the role which micro-organisms play is complex and numerous mechanisms have been suggested to explain their involvement in the corrosion process. Rarely does their intrusion into the corrosion process involve any new mechanism, but by facilitating the formation of differential aeration and concentration cells, or the production of aggressive metabolites, or by their ability to accelerate an existing corrosion process, the organisms can promote serious corrosion problems. The economic significance of this field has been highlighted by a number of surveys.

The organisms responsible for the majority of these problems are the sulphate-reducing bacteria. Not only do they produce an aggressive metabolite but their metabolism allows them to utilize cathodic hydrogen and thus accelerate the corrosion process. Extensive research has been carried out to clarify the mechanisms by which these bacteria function in the corrosion process but a complete answer has not yet been obtained. Evidence in support of the classical theory still appears to be valid, but alternative suggestions provide interpretations for some of the experimental observations. Overall the role of the sulphate-reducing bacteria seems to be associated with cathodic depolarization, although in certain circumstances the effects of sulphide metabolic products may be of greater importance. The cathodic characteristics of iron sulphide have been known for some time but only recently has an explanation been given for the mechanisms involving sulphides and indeed elemental sulphur in the corrosion process. Again, in order to interpret field observations, there is a need to perform more experiments in near-natural systems. To date few laboratory studies have been able to simulate field conditions, particularly the changes in redox potential which must occur during the corrosion process. Similar considerations also apply to growth and nutritional studies.

Aerobic conditions involving biofouling and subsequent microbial corrosion are now being studied more extensively and the mechanism by which bacteria adhere to metal surfaces has already received extensive study.[114-116] The influence of temperature under both anaerobic and aerobic conditions is also a neglected area which requires further study. Temperature effects are seen to be particularly important where heat transfer conditions exist as in fouled heat exchangers and on hot oil riser pipes.

From a practical engineer's point of view, positive steps can be taken to avoid microbial corrosion problems. The review has discussed some of the simple qualitative assessments which can be made to determine the risk of microbial corrosion in particular environments, and means for avoiding microbial problems by the selection of materials, barrier systems or modification of the environment. For internal problems in process systems, control of microbial corrosion by chemical means is the favoured solution. Professional guidance is generally required in the selection of appropriate materials from a long and ever-increasing list, but this can frequently be provided as part of an overall water treatment package.

Microbial corrosion is a field which is just beginning to enjoy a renaissance. Luty[117] considers the sudden increase in awareness to be due to an increase in the rate of publication, particularly related to biofouling aspects of the problem. There are a number of unanswered questions which have recently increased in importance, for example in the North Sea oil and gas industry, and in the water-handling field where changes from the use of strong inhibitors to careful scale-control procedures, together with a drive to save water and energy, have led to an increased incidence of this type of corrosion problem. There are however pitfalls in interpretation for the unwary, for it is quite possible to detect microbial presence in many corrosion problems where they make no serious impact on the severity of the problems, which would have occurred even in the absence of microbial involvement. Nevertheless, greater awareness can only be generally beneficial, for only by wider acceptance of the 'state of the art' consensus will designers and plant operators automatically consider the design features, operating procedures and system monitoring which are important if microbial problems are to be contained and controlled in industrial situations.

ACKNOWLEDGEMENT

I am grateful to my colleague Dr G. P. Rothwell for his encouragement and assistance during the preparation of this review.

REFERENCES

1. U. R. Evans, *Corrosion and oxidation of metals*, Edward Arnold, London (1960).
2. L. L. Shrier (ed.), *Corrosion*, Vol. 1 (2nd edn). Newnes Butterworth, London (1976).
3. D. S. Wakerley, Microbial corrosion in UK industry; a preliminary survey of the problem. *Chem. and Ind.*, 656 (1979).
4. J. D. A. Miller (ed.), *Microbial aspects of metallurgy*. MTP, Aylesbury (1971).
5. G. H. Booth, *Microbial corrosion*, M & B Monograph CE/1. Mills & Boon, London (1971).
6. W. P. Iverson, Biological corrosion, in *Advances in corrosion science and technology*, Vol. 2, eds M. G. Fontana and R. W. Staehle, 1. Plenum Press, London (1972).
7. J. D. A. Miller and R. A. King, Biodeterioration of metals, in *Microbial aspects of the deterioration of material*, eds D. W. Lovelock and R. J. Gilbert, 83. Acad. Press, London (1975).

8. J. Knox, The role of government in international co-operative research, in *Biodeterioration of materials*, eds A. H. Walters and J. J. Elphick, 1. Elsevier, London (1968).

9. W. H. S. Vernon, Metallic corrosion and conservation. *Conservation of natural resources*. 105. Inst. Civil Eng., London (1956).

10. B. E. Purkiss, Corrosion in industrial situations by mixed microbial floras, in *Microbial aspects of metallurgy*, ed. J. D. A. Miller, 107, MTP, Aylesbury (1971).

11. N. W. Le Roux and D. S. Wakerley, Microbial corrosion—a preliminary survey of the problem in UK industry. CR1505(ME), Warren Spring Laboratory (1978).

12. A. M. Pritchard, Heat exchanger fouling in British industry. *Fouling Prevention Res. Digest*, 1, iv (1979).

13. J. H. Garrett, *The action of water on lead*. H. K. Lewis, London (1891).

14. R. H. Gaines, Bacterial activity as a corrosion influence in the soil. *J. Eng. Ind. Chemistry*, 2, 128 (1910).

15. C. A. H. von Wolzogen Kühr and L. S. van der Vlugt, De grafiteering van Gietijzer als electrobiochemich Proces in anaerobe Gronden. *Water* (den Haag) 18 (16), 147 (1934).

16. D. Ellis, *Iron bacteria*, Methuen, London (1919).

17. E. C. Harder, *Iron depositing bacteria and their geologic relations*, Govt Printing Office, Washington DC (1919).

18. R. F. Hadley, The influence of *Sporovibrio desulfuricans* on the current and potential behavior of corroding iron. NBS Soil Corrosion Conference, St. Louis (1943).

19. K. Duchon and L.B. Miller, The effects of chemical agents on iron bacteria. *Paper Trade J.*, 124 (4), 47 (1948).

20. W. Szybalski and F. Olsen, Aerobic microbial corrosion of water pipes. *Corrosion*, 6, 405 (1949).

21. K. R. Butlin, M. E. Adams and M. Thomas, Sulphate-reducing bacteria and internal corrosion of ferrous pipes conveying water, *Nature*, 163, 26 (1949).

22. J. N. Wanklyn and C. J. P. Spruit, Influence of sulphate-reducing bacteria on the corrosion potential of iron, *Nature*, 169, 929 (1952).

23. H. H. Uhlig, *Corrosion handbook* (4th edn), John Wiley, New York (1953).

24. E. C. Hills, Microbial infections in relation to corrosion during metal machining and deformation, in *Microbial aspects of metallurgy*, ed. J. D. A. Miller, 129. MTP, Aylesbury (1971).

25. D. G. Parberry, The role of *Cladosporium resinae* in the corrosion of aluminium alloys. *Int. Biodeterioration Bulletin*, 4 (2), 79 (1968).

26. J. J. Elphick, Microbial corrosion in aircraft fuel systems, in *Microbial aspects of metallurgy*, ed. J. D. A. Miller, 157. MTP, Aylesbury (1971).

27. S. H. Waksman, *Principles of soil microbiology*, Williams & Wilkins, Baltimore (1927).

28. R. L. Starkey, Relations in micro-organisms to transformation of sulphur in soil, *Soil Science*, 70, 55 (1950).

29. E. S. Kempner, Acid production by *Thiobacillus thiooxidans*, *J. Bact.*, 92, 1842 (1966).

30. C. D. Parker, Mechanics of corrosion of concrete sewers by hydrogen sulphide, *Sewage Ind. Wastes*, **23**, 1477 (1951).
31. J. Pochon and C. Jaton, Facteurs biologiques de l'alteration des pierres, in *Biodeterioration of materials*, eds A. H. Walters and J. J. Elphick, 258. Elsevier, London (1968).
32. R. G. Ainsworth, J. Ridgway and R. D. Gwilliam, Corrosion products and deposits in iron mains. Water distribution systems—maintenance of water quality and pipeline integrity, Oxford 1978, Paper 8. Water Research Centre, Medmenham (1978).
33. E. C. Pringsheim, Iron bacteria, *Biol. Rev. Cambridge Phil. Soc.*, **24** (2), 200 (1949).
34. J. S. Stokes, Studies on the filamentous-sheathed iron bacterium Sphaerotilus natans. *J. Bact.*, **67** (3) 278 (1954).
35. H. D. Peck, Symposium on metabolism of inorganic compounds: V. Comparative metabolism of inorganic sulphur compounds in microorganisms. *Bacteriol. Rev.*, **26**, 67 (1962).
36. J. R. Postgate, *The sulphate-reducing bacteria.* CUP, Cambridge (1979).
37. J. D. A. Miller and A. K. Tiller, Microbial corrosion of buried and immersed metal, in *Microbial aspects of metallurgy*, ed. J. D. A. Miller, 61. MTP, Aylesbury (1971).
38. A. M. Saleh, Differences in the resistance of sulphate-reducing bacteria to inhibitors, *J. Gen. Microbiol.*, **37**, 113 (1964).
39. G. H. Booth, A. K. Tiller and F. Wormwell, A laboratory study of well-preserved ancient iron nails from apparently corrosive soils. *Corrosion Sci.*, **2**, 197 (1962).
40. D. J. Crombie, G. J. Moody and J. D. R. Thomas, Corrosion of iron by sulphate-reducing bacteria, *Chem. and Ind.*, **21**, 500 (1980).
41. D. K. B. Thistlewayte, *Control of sulphides in sewage systems*, Butterworths, London (1972).
42. A. M. Douglas, Corrosion investigation at Burton-on-Trent, *J. Chartered Mun. Eng.*, **191**, 130 (1964).
43. N. W. Le Roux, Mineral attack by microbiological processes, in *Microbial aspects of metallurgy*, ed. J. D. A. Miller, 173. MTP, Aylesbury (1971).
44. A. W. Fletcher, Copper recovery from low grade ore by bacterial leaching, in *Microbial aspects of metallurgy*, ed. J. D. A. Miller, 183. MTP, Aylesbury (1971).
45. W. P. Iverson, A possible role for sulphate-reducers in the corrosion of aluminium alloys, *Electrochem. Tech.*, **5** (3–4), 77 (1967).
46. G. C. Blanchard and C. R. Goucher, Aluminium corrosion processes in microbial cultures, *Electrochem. Tech.*, **5** (3–4), 79 (1967).
47. A. K. Tiller and G. H. Booth, Anaerobic corrosion of aluminium by sulphate-reducing bacteria, *Corros. Sci.*, **8**, 549 (1968).
48. W. J. Copenhagen, The pathology of metals: corrosion of steel and Alclad parts by a fungus, *Met. Ind.*, **77**, 137 (1950).
49. O. H. Calderon, E. E. Stratfeld and C. B. Coleman, Metal–organic acid corrosion and some mechanisms associated with these processes, in *Biodeterioration of materials*, eds A. H. Walters and J. J. Elphick, 356. Elsevier, New York (1968).

50. J. Bunker, Microbiological experiments in anaerobic corrosion, *J. Soc. Chem. Ind.*, **58**, 93 (1939).

51. C. J. P. Spruit and J. N. Wanklyn, Influence of sulphate-reducing bacteria on the potential of iron, *Nature*, **169**, 928 (1952).

52. J. Horvath, Contributions to the mechanisms of anaerobic microbiological corrosion, *Acta Chem. Acad. Sci. Hung.*, **25**, 65 (1960).

53. G. H. Booth and F. Wormwell, Corrosion of mild steel by sulphate-reducing bacteria: effect of different strains of organisms, *1st Int. Cong. on Metallic Corrosion*, 1961, 341. Butterworth, London (1962).

54. G. H. Booth, L. Elford and D. S. Wakerley, Corrosion of mild steel by sulphate-reducing bacteria: an alternative mechanism, *Br. Corr. J.*, **3**, 242 (1968).

55. R. A. King and J. D. A. Miller, Corrosion by the sulphate-reducing bacteria, *Nature*, **233**, 491 (1971).

56. D. D. Mara and D. S. A. Williams, The mechanism of sulphide corrosion by sulphate-reducing bacteria, in *Biodeterioration of materials*, Vol. 2, eds A. H. Walters and E. H. Hueck-van der Plas, Applied Science, Barking (1972).

57. J. S. Smith and J. D. A. Miller, Nature of sulphides and their corrosive effect on ferrous metals—a review. *Br. Corr. J.*, **10**, 136 (1975).

58. F. H. Meyer, O. L. Riggs, R. L. McGlasson and J. D. Sudbury, Corrosion products of mild steel in hydrogen sulfide environments, *Corrosion*, **14**, 109t (1958).

59. W. F. Rogers and J. A. Rowe, *Proc. 4th World Pet. Congr. Rome*, Vol. II, 479 (1955).

60. J. L. Battle, Corrosion of casing in oil and gas wells, *Corrosion*, **9**, 313 (1953).

61. L. R. Sheppard, discussion on Corrosion of steel by hydrogen sulphide–air mixtures, by D. C. Bond and G. A. Marsh, *Corrosion*, **6**, 27 (1950).

62. W. R. Scott, Bacterial corrosion in a waterflood system. *Mat. Protection*, **4** (2), 57 (1965).

63. G. A. Trautenberg, Sulphate reduction in bacterial corrosion, *Mat. Protection*, **3** (2), 30 (1964).

64. E. Schaschl, Elemental sulphur as a corrodent in deaerated neutral aqueous environments. *Mat. Performance*, **19** (7), 9 (1980).

65. J. D. A. Miller and D. S. Wakerley, Growth of sulphate-reducing bacteria by fumarate dismutation, *J. Gen. Microbiol.*, **43**, 101 (1966).

66. M. Stephenson and L. H. Stickland, Hydrogenase II, The reduction of sulphate to sulphide by molecular hydrogen, *Biochem. J.*, **25**, 215 (1931).

67. G. H. Booth and A. K. Tiller, Polarization studies of mild steel in cultures of sulphate-reducing bacteria. *Trans. Faraday Soc.*, **56**, 1689 (1960).

68. A. K. Tiller and G. H. Booth, Polarization studies of mild steel in cultures of sulphate-reducing bacteria: Part 2, Thermophilic organisms. *Trans. Faraday Soc.*, **58**, 110 (1962).

69. G. H. Booth and A. K. Tiller, Polarization studies of mild steel in cultures of sulphate-reducing bacteria: Part 3, Halophilic organisms. *Trans. Faraday Soc.*, **58**, 2510 (1962).

70. J. Horvath and M. Solti, Beitrag zum Mechanismus der anaerobischen mikrobiologischen Korrosion der Metalls im Boden. *Werkstoffe und Korrosion*, **10**, 624 (1959).

71. W. P. Iverson, Direct evidence for the cathodic depolarization theory of bacterial corrosion, *Science*, **151** (3713), 986 (1966).
72. G. H. Booth and A. K. Tiller, Cathodic characteristics of mild steel in suspensions of sulphate-reducing bacteria, *Corrosion Sci.*, **8**, 583 (1968).
73. G. H. Booth, P. M. Shinn and D. S. Wakerley, The influence of various strains of actively growing sulphate-reducing bacteria on the anaerobic corrosion of mild steel. *Congrès Int. de la Corrosion Marine et des Salissures*, Cannes 1964, 363. Eds. CREO, Paris (1965).
74. G. H. Booth, A. W. Cooper and P. M. Cooper, Rates of microbial corrosion in continuous cultures, *Chem. and Ind.*, **86**, 2084 (1967).
75. R. Stumper, La corrosion du fer dans la présence du sulphur de fer, *Compt. Rend.*, **176**, 1316 (1923).
76. R. S. Treseder, San Francisco Western Region NACE meeting (1955).
77. T. W. Farrer and F. Wormwell, Corrosion of iron and steel by aqueous suspensions of sulphur. *Chem. and Ind.*, **72**, 106 (1953).
78. G. H. Booth, J. A. Robb and D. S. Wakerley, The influence of ferrous ions on the anaerobic corrosion of mild steel by actively growing cultures of sulphate-reducing bacteria. *3rd Int. Congress on Metallic Corrosion*, Moscow 1967. Vol. 2, Sect. 5, 542.
79. R. A. King, J. D. A. Miller and D. S. Wakerley, Corrosion of mild steel in cultures of sulphate-reducing bacteria: effect of changing soluble iron concentration during growth, *Br. Corr. J.*, **8**, 89 (1973).
80. G. H. Booth, P. M. Cooper and D. S. Wakerley, Corrosion of mild steel by actively growing cultures of sulphate-reducing bacteria: the influence of ferrous iron, *Br. Corr. J.*, **1**, 345 (1966).
81. R. A. King, J. D. A. Miller and J. S. Smith, Corrosion of mild steel by iron sulphides. *Br. Corr. J.*, **8**, 137 (1973).
82. R. A. King and J. D. A. Miller, Corrosion of ferrous metals by bacterially produced iron sulphide and its control by cathodic protection. *1st Int. Conf. on the Internal and External Protection of Pipes*, Durham, 1975. Paper F2–9. BHRA, Cranfield (1975).
83. D. T. Rickard, The microbiological formation of iron sulphides. Stockholm contr. to *Geology*, **20**, 67 (1969).
84. C. G. Munger, Deep pitting corrosion in sour crude oil tankers, *Mat. Performance*, **16** (3), 17 (1976).
85. K. R. Butlin and W. H. J. Vernon, Underground corrosion of metals: causes and prevention, *J. Inst. Water Eng.*, **3**, 627 (1949).
86. R. L. Starkey and K. M. Wright, Anaerobic corrosion of iron in soil, Amer. Gas Ass., New York (1945).
87. M. Romanoff, Underground corrosion, Circular 579, 227. NBS, Washington DC (1959).
88. R. F. Stratfull, A new test for estimating soil corrosivity based on investigations of metal highway culverts, *Corrosion*, **17**, 493t (1961).
89. G. H. Booth, A. W. Cooper and D. S. Wakerley, Criteria of soil aggressiveness towards buried metals: I. Experimental methods, *Br. Corr. J.*, **2**, 104 (1967).
90. G. H. Booth, A. W. Cooper and P. M. Cooper. Criteria of soil aggressiveness towards buried metals: II. Assessment of various soils, *Br. Corr. J.*, **2**, 109 (1967).

91. G. H. Booth, A. W. Cooper and A. K. Tiller. Criteria of soil aggressiveness towards buried metals: III. Verification of predicted behaviour of selected soils, *Br. Corr. J.*, **2**, 116 (1967).

92. D. D. Mara and D. J. A. Williams. The evaluation of media used to enumerate sulphate-reducing bacteria, *J. Appl. Bact.*, **33**, 543 (1970).

93. R. A. King, Prediction of corrosiveness of sea-bed sediments. *Mat. Performance*, **19** (1), 39 (1980).

94. J. Morley, A review of the underground corrosion of steel piling. BSC Report T/CS/1114/1/78/C. BSC, Teesside Laboratories (1978).

95. W. J. Schwerdtfeger and M. Romanoff, NBS papers on underground corrosion of steel piling. US Dept. of Commerce, Washington DC (1962–1971).

96. G. R. Eadie, The durability of steel pilings in soil. Broken Hill Pty, Melbourne Res. Labs (1978).

97. Y. Osaki, Corrosion of steel piles in soil deposits. 8th Int. Congress on Metallic Corrosion, Tokyo, 1972.

98. K. Tungesvik, J. Moum and K. P. Fiseher, Investigations of corrosion rates on steel piles in Norwegian marine sediments. *Proc. 7th Scandinavian Corr. Conf.*, Trondheim, 1975, p. 487.

99. G. H. Booth, A. W. Cooper and A. K. Tiller, Corrosion of mild steel in the tidal waters of the Thames estuary: I. Results of six months and one year immersion, *J. App. Chem.*, **13**, 211 (1963).

100. G. H. Booth, A. W. Cooper and A. K. Tiller, Corrosion of mild steel in the tidal waters of the Thames estuary: II. Results of two years immersion, *J. App. Chem.*, **15**, 250 (1965).

101. G. H. Booth, A. W. Cooper and A. K. Tiller, Corrosion of mild steel in the tidal waters of the Thames estuary: III. Results of three and five years immersion, *Br. Corr. J.*, **2**, 222 (1967).

102. G. H. Booth, A. W. Cooper and A. K. Tiller, Corrosion of mild steel in the tidal waters of the Afon Dyfi and a comparison with the River Thames, *Br. Corr. J.*, **2**, 21 (1967).

103. B. E. Purkiss and R. W. N. Cameron. Pipeline corrosion inhibited by a bactericide. *Gas Journal Supplement* 3–8, (Sept–Oct. 1965).

104. 1st Int. Conf. on the Internal and External Protection of Pipes, Durham, 1975. BHRA, Cranfield (1975).

105. 2nd Int. Conf. on the Internal and External Protection of Pipes, Canterbury, 1977. BHRA, Cranfield (1978).

106. 3rd Int. Conf. on the Internal and External Protection of Pipes, London, 1979. BHRA, Cranfield (1979).

107. Protections for tubular steel products. BSC Tubes Div., Corby (1978).

108. E. S. Pankhurst, Bacteria, barnacles and biomass. Communication 1070, Inst. of Gas Engineers (1978).

109. J. H. Morgan, *Cathodic protection*, Leonard Hill, London (1959).

110. J. Horvath and M. Novak, Potential/pH equilibrium diagrams of some M–S–H_2O ternary systems and their interpretation from the point of view of metallic corrosion. *Corrosion Sci.*, **8**, 583 (1968).

111. B. Creedon, Biofouling in cooling water plant. Conference on Fouling—Science or Art, Guildford, 1979.

112. J. W. McCoy, *The chemical treatment of cooling water*. Chemical Publishing, New York (1974).
113. A. M. Saleh, R. Macpherson and J. D. A. Miller, The effect of inhibitors on sulphate-reducing bacteria: A compilation, *J. Appl. Bact.*, **27**, 281 (1964).
114. J. W. Costerton and G. G. Gessey, How bacteria stick. *Sci. Amer.*, **238**(1), 86 (1978).
115. W. G. Characklis, Microbial fouling, a process analysis. Int. Conf. on Fouling of Heat Transfer Equipment, Troy, NY (1979).
116. J. W. Costerton, Mechanism of primary fouling of submerged surfaces by bacteria. Int. Conf. on Fouling of Heat Transfer Equipment, Troy, NY (1979).
117. R. Luty, Microbiological corrosion. Paper 39, NACE Corrosion Conference '80, Chicago (1980).

Chapter 4

LOCALIZED CORROSION

D. A. JONES

Department of Chemical and Metallurgical Engineering,
University of Nevada,
USA

1. INTRODUCTION

The purpose of this chapter is to review the literature of recent years and summarize current progress on the mechanisms of various forms of localized corrosion. These forms are taken here to include galvanic (two-metal) corrosion, dealloying, intergranular attack (IGA), erosion corrosion, and finally crevice and pitting corrosion. Galvanic corrosion is treated initially because there are galvanic elements in all of the other localized forms of corrosion. The greatest bulk of research has been devoted to the last-treated subject, crevice and pitting corrosion. The intervening subjects are discussed toward the beginning of this chapter in the hope that further research activity may be stimulated in these no less important topics.

Electrochemical techniques continue to grow in their usefulness in elucidating mechanisms and providing accelerated test methods for detecting and testing resistance to localized corrosion. These are discussed where appropriate.

Mechanical interactions with corrosion including stress corrosion and corrosion fatigue will be discussed in other chapters of the present volume. Also falling into this category are certain cavitation phenomena (see, for example, C. M. Preese, ed., *Erosion: Treatise on Materials Science and Technology*, Vol. 16, Academic Press, New York, (1979)) in which the ambient fluid serves primarily as a pressure transfer medium with minimal effects of corrosion and significant local plastic deformation of the metal surface.

The localized forms of corrosion have always assumed great importance because it has been difficult to predict service life in the presence of local attack. Localized corrosion is especially troublesome for the corrosion-resistant alloys which have served reliably in other similar applications. Current technology frequently imposes more severe conditions of temperature, pressure, and design than hitherto and unexpected localized corrosion is often the result. A recent example from the energy generation industry is shown in Fig. 1.[1] A crevice is present

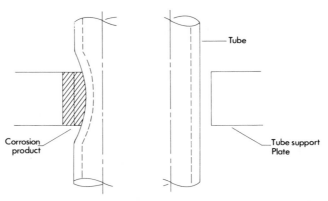

FIG. 1. Denting in the crevice between tube and tube support plate in the steam generator of a PWR.[1]

between the Inconel heat exchanger tubes and the carbon steel tube supports in a steam generator of a pressurized water nuclear reactor (PWR). Under certain secondary water chemistry conditions, corrosion products grow in the crevice, distorting the tubes and cracking the tube supports. The mechanism of denting has not been well defined, but a brief review of current knowledge on the subject is given in the Appendix because of practical importance and current urgency.

No effort has been made to be complete historically in the interest of time and space. Literature of the past 10–12 years is emphasized, although older work is mentioned occasionally where it seems to amplify the discussion. The interested reader should refer to the literature cited, especially recent reviews when available, for more complete references to the important older work on which much of our current understanding is based.

The author has attempted whenever possible to draw the many divergent elements of the literature into a coherent direction, and the most important recent works have been cited in support of the apparently most logical and conclusive interpretations. In view of the voluminous literature, it has been found prohibitive to cite every work pertaining to any particular aspect of the subject. These judgements and interpretations are inescapably subjective. The author offers his apologies to investigators, past and present, known and unknown, for any serious omissions.

2. GALVANIC CORROSION

We will see in subsequent sections of this chapter that galvanic corrosion plays an important role in nearly every form of localized corrosion. It is well known that the corrosion of a metal is characterized by a corrosion potential, E_{corr}, and a corrosion rate or current density, i_{corr}, in a given corrosive electrolyte.[2,3] A listing of corrosion potentials for various metals and alloys constitutes a galvanic series for the particular corrosive electrolyte in which the potentials are measured. An alloy which is active (negative) in a galvanic series will often have its corrosion rate increased significantly when coupled to an alloy with a corrosion potential which is more noble (positive) in the galvanic series. However, recent work [4,5,6] has emphasized that the potential difference between a pair of alloys gives no quantitative prediction of galvanic corrosion. Polarization at the anode and/or the cathode will limit the actual galvanic corrosion rate, which is measured by the current, I_{GSC}, at the galvanic short circuit with essentially zero resistance between anode and cathode.[7]

The importance of polarization in galvanic couples is shown in Fig. 2. Polarization at the cathode, anode, or both may control the galvanic current, I_{GSC}, which can flow in the couple. Polarization curves for anode and cathode may be determined by step-wise change in the potential or current.[7] Figure 3 shows that the cathode polarization curve increases in potential with decreasing current, while the anode polarization curve decreases in potential. These experimental 'Evans diagrams' give substantial insight into the mechanism of galvanic corrosion and the best means by which it may be controlled.

The effect of surface area ratio of cathode to anode may also be predicted from the polarization diagrams. Figure 4 shows schematically the effect of increasing the cathode area in a galvanic couple and

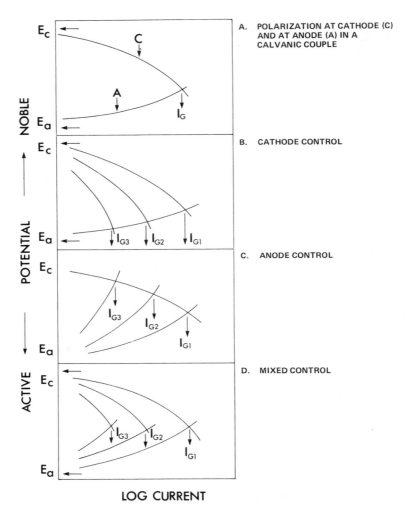

FIG. 2. Effect of polarization on galvanic behavior in dissimilar metal couples.

indicates that control changes from cathode to anode control at high cathode:anode surface area ratios. The predictions shown in Figs 2 and 4 have been fulfilled experimentally for brass–steel couples in salt water.[8] The area ratio can have a variety of effects on the behavior of galvanic couples and some general cases are described analytically by Mansfeld.[9]

The usual galvanic series will predict which half of a couple will act as

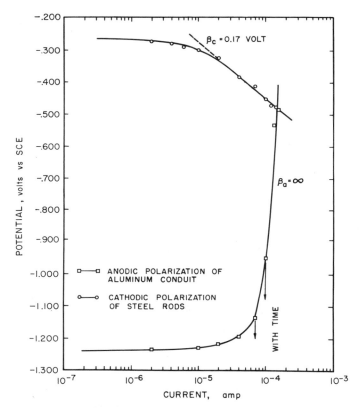

FIG. 3. Polarization diagram showing anode and cathode polarization in a galvanic couple of aluminum and stainless steel in cement mortar with 2% CaCl₂.[7] By courtesy of The Electrochemical Society.

anode and therefore be attacked preferentially.[4] However, the galvanic current, I_{GSC}, is needed to predict the actual extent of galvanic attack and Mansfeld et al.,[4] and Walker[6] have ranked various couples according to I_{GSC}. Such a modified galvanic series will be complex because each couple has a unique value for I_{GSC} for each corrosive environment and area ratio but Table 1 indicates results for a number of galvanic couples studied by Walker[6] in 3.5% NaCl.

Also shown in Table 1 are the couple potentials for equal surface areas in 3.5% NaCl. Mansfeld et al.[4] have shown that when polarization at the anode is rather small the local action or self-corrosion at the anode is

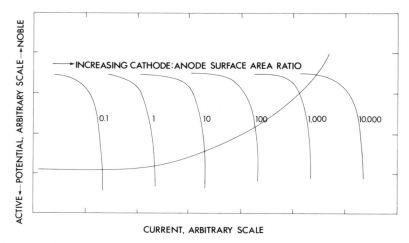

FIG. 4. Effect of surface area ratio on the couple potential and couple current in a galvanic couple (schematic).

still a major part of the total anode corrosion rate, and I_{GSC} is not a good measure of the absolute dissolution rate of the anode. Inspection of Table 1 shows that quite often the couple potential is rather near the uncoupled potential, E_{corr}, of the anode and Walker[6] showed that the anode dissolution rate is not exactly matched by I_{GSC}, although the increased corrosion rate due to galvanic coupling, $I_{total} - I_{corr}$, was near I_{GSC} in most cases. I_{total} and I_{corr} are the weight-loss-determined corrosion rates for the coupled and freely corroding anode specimens respectively, converted to current for comparison and Walker's results confirm the relationship given by Mansfeld et al.,[4] $I_{GSC} \cong I_{total} - I_{corr}$, for low anode polarization. I_{GSC} will still be a good measure of total corrosion if $I_{total} \gg I_{corr}$.

Measurement of I_{GSC} requires that the anode and cathode be polarized to very nearly equal potentials, as would be the case for short-circuit conditions with no intervening resistance. A normal ammeter has a finite resistance, R, and measurement of galvanic current, I_G, is lower than I_{GSC} for short-circuit conditions. The anode and cathode potentials, E_a and E_c respectively, are separated by an amount $I_G R$. This ohmic separation is overcome conventionally with the galvanostatic zero resistance ammeter (ZRA) shown in Fig. 5. A battery or dc power supply is used to polarize the electrodes to the same potential and the current required to achieve this is I_{GSC}.

TABLE 1

GALVANIC SERIES BASED ON GALVANIC SHORT CIRCUIT CURRENT AND COUPLE POTENTIAL. FOR EACH COUPLE, UPPER ENTRY IS I_{GSC} IN μA/CM^2 AND LOWER ENTRY IS COUPLE POTENTIAL IN mV v. SCE, NEGATIVE. EQUAL AREAS OF ANODE AND CATHODE DATA COMPILED FROM REF. 7. 3.5% NaCl AERATED.

Anode	$-E_{corr}$ (mV/SCE)	Nickel	301 Stainless	201 Stainless	430 SS	Brass	409 SS	Solder	Steel	Aluminized Steel	Aluminum	Zincro Metal
Nickel	37											
301 SS	138											
201 SS	138											
430 SS	170											
Brass	261											
409 SS	279											
Solder	538					19.3 / 482						
Steel	690	15.3 / 690	8.5 / 704	10.9 / 676		24.2 / 690		12.0 / 645				
Aluminized Steel	723								18.8 / 727			
Aluminum	725	16.0 / 722						13.5 / 719	33.8 / 725	−2.0 / 739		
Zincro-Metal	859									3.3 / 824	14.6 / 753	
Galv. Steel	1080										76.7 / 1070	3.1 / 1070

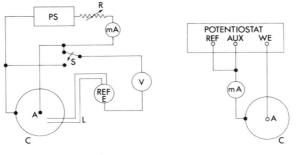

A. GALVANOSTATIC ZRA   B. POTENTIOSTATIC ZRA

FIG. 5. Zero resistance ammeter (ZRA) circuits in which there is, in effect, zero resistance between anode and cathode of a galvanic couple.[7] PS: power supply; A and C: anode and cathode; MA: milliammeter; V: voltmeter; L: salt bridge; REF.E: reference electrode; REF, AUX and WE: reference, auxiliary, and working electrode terminals of potentiostat.

Current at the galvanic short circuit, I_{GSC}, varies with time and the galvanostatic ZRA requires continuous adjustment to obtain I_{GSC} as a function of time but Jones[7] and Devay et al.[10] have shown that a potentiostat can be used to obtain a continuous record of I_{GSC} with the very simple circuit shown in Fig. 5B. The potentiostat is adjusted to control $\Delta E = 0$ between the two halves of the galvanic couple and the potentiostatic current is then equal to I_{GSC}. Others[11,12] have since proposed operational amplifier circuits to serve the same function, but these have not gained wide acceptance,[13] probably because of the ready availability of potentiostats in most laboratories.

3. DEALLOYING

Heidersbach and Verink[14,15] have conducted a detailed investigation of dezincification of α and $\alpha + \beta$ brasses. They conclude that two mechanisms may operate separately or simultaneously: (a) selective dissolution of the more active zinc; and (b) total dissolution followed by redeposition of dissolved copper. This conclusion is supported by more recent work of Natarajan et al.[16]

Langenegger and Callaghan[17] have demonstrated that dezincification occurs under oxidizing conditions, imposed chemically or electrochemically, in chloride solutions of varying concentration and pH. The 'potential shift' caused by adding chemical oxidizers was found to correlate with the rate of surface penetration by dezincification, so that the susceptibility to dezincification can be predicted from the difference in potentials between that in the oxidizing solution and that in a solution of equivalent pH and chloride concentration but with no oxidizers. Advanced dezincification could be immediately stopped by removing the oxidizing conditions. This led to the conclusion that a build-up and redeposition of Cu^{2+} in the dezincified structure is not a primary factor in the mechanism. However, redeposition of Cu^{2+} could be a secondary process which occurs subsequent to the controlling process of selective dissolution.

The equivalent effects of chemical and electrochemical oxidizing conditions have led to the belief that electrochemical techniques may be of promise in developing accelerated tests for dealloying and some attempts at such have been reported recently. Costas[18] applied a potentiostatic anodic potential to brass in simulated Colorado River water for a period of 24 h and inspected the structure metallographically for evidence of dezincification but the potentiostatic current during exposure was not a good indicator, especially at low levels, of dezincification. Lucey[19] exposed brass specimens to a saturated solution of $CuCl_2$ with $ZnCl_2$ while the potential of the brass was measured against that of pure Cu. A large potential developed during the 30 min test duration indicated susceptibility to dezincification. Maahn and Blum[20] found that when brass specimens were potentiostatically controlled at the free corrosion potential of copper in a solution of 0.5 M NaCl, 0.22 M sodium acetate, and 0.68 M acetic acid with pH = 4.2 for a period of 24 h, the steady state potentiostatic current gave a ranking of the tendency for dezincification of various brasses. Dillon[21] compared the Lucey[19] and Maahn-Blum[20] tests with conventional two-exposure tests to oxidizing $CuCl_2$ solutions and reported good correlation in both cases. Nevertheless, there appears to be continuing reluctance to use electrochemical tests and conventional $CuCl_2$ exposures are still favored, although increasing the temperature to $75\,^{\circ}C$ has decreased exposure time to 24 h.[22]

The polarization tests discussed above measure the kinetics of dezincification. There has been some success in using potential–pH diagrams to predict dealloying in copper–nickel[23] and copper–zinc[24] alloys, but predictions for Fe–Ni–Cr alloys[25] have not been fulfilled.[26]

4. INTERGRANULAR ATTACK (IGA)

Intergranular attack, IGA, of austenitic stainless steel has long been explained by the chromium depletion theory, in which carbon diffuses to the grain boundaries and reacts with chromium to form chromium carbides, thereby depleting the adjacent areas of chromium. Since stainless steels depend on chromium for corrosion resistance, the grain boundary areas become less resistant to corrosion and more susceptible to localized attack. A direct proof of the chromium depletion theory has been provided by Joshi and Stein,[27] who observed chromium depletion and nickel enrichment at the intergranular fracture surfaces of sensitized austenitic stainless steel using Auger electron spectroscopy. Latanision et al. [28] have shown direct evidence of grain boundary depletion of chromium in Inconel using scanning transmission electron microscopy (STEM).

There has been renewed interest recently in the duplex stainless steels,[29] in which ferrite is retained in a metastable state by rapid quenching from above 1050 °C, where both the α and γ phases are stable. Recent work[30-32] has shown that chromium diffuses faster in the α phase than in austenite at sensitizing temperatures and the chromium carbides form preferentially at the α–γ boundaries and, if carbon in the austenite is exhausted, prevent carbide precipitation at the austenite grain boundaries. There is a narrow chromium depleted band in γ adjacent to the α–γ boundaries, due to a smaller but finite rate of diffusivity of chromium in austenite. However, this small depleted zone is easily replenished in chromium by diffusion during service stress-relief annealing of typically 600 °C for 24–48 h.[31,32] The ferrite is unstable at the annealing temperature and decomposes after about 2 h leaving behind a cellular structure of carbide in austenite. The austenite, although somewhat depleted in chromium, is still resistant to general attack and IGA.[31]

France and Greene[33] used electrochemical polarization to differentiate conditions which can cause IGA in sensitized alloys, producing maps of potential v. acid concentration showing regions where IGA would be found. Figure 6 gives a sample of their data from specimens inspected for IGA after 24 h at each potential. Although France and Greene cautioned that much longer exposure times would be required before service conditions could be predicted, Streicher[34] argued that sensitized alloys should always be expected to fail by IGA. France and Greene[35] replied by citing sensitized alloy service exposures as long as eight months with

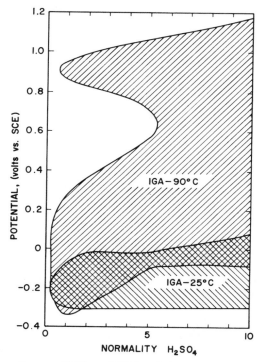

F<small>IG</small>. 6. Regions of susceptibility of 18-8 stainless steel to intergranular attack (IGA) on a map of potential v. acid concentration.[33]

no evidence of IGA and there clearly are some applications where marginally sensitized material is useful and only heavily sensitized material is unsuitable.[36]

Electrochemical techniques have enjoyed success in accelerated testing for sensitization to IGA. Cihal et al.[37] reported a potentiodynamic repassivation method to detect sensitization in austenitic stainless steel and potentiodynamic cathodic potential sweep procedures from passive potentials have been used for specimens in 0.5 M H_2SO_4 $+0.01$ M KSCN.[36,38] The specimen is held for a few minutes at a passive potential in this solution, and a cathodic potential sweep is conducted to below E_{corr}, the stable corrosion potential. Data due to Clarke et al.[36] are shown in Fig. 7 for specimens of austenitic stainless steel sensitized for 4 h at increasing temperatures from 900 °F to 1200 °F, the degree of sensitization being related to the area under the 'reactivation' current peak during the

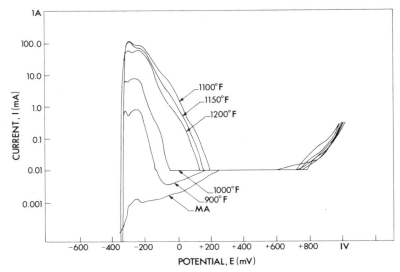

FIG. 7. Potentiokinetic reactivation curves showing effect of sensitization temperature (4 h) on Type 304 stainless steel.[36] By courtesy of the National Association of Corrosion Engineers.

cathodic potential scan. The non-sensitized, mill-annealed specimen is able to maintain passivity throughout the reactivation scan, but sensitized materials show reactivation in the active potential region due to breakdown of the passive film at the chromium-depleted grain boundaries. The degree of breakdown is apparently proportional to the amount and distribution of chromium carbides at the grain boundaries.

A standardized, electrochemical potentiokinetic repassivation (EPR) procedure for detecting sensitivity to IGA has been proposed[36] with the conditions summarized in Table 2. A coulometer is included in the circuit to measure the charge Q passed under the reactivation peak (Fig. 7) and since the current peak will be larger for larger sample surface or grain boundary areas (smaller grain size), Q is adjusted to a value P_a for unit exposed grain boundary area (Table 2). The EPR procedure gives a sensitive indication of the degree of sensitization especially at low levels of sensitization. Any particular value of P_a can be selected as a tolerable level of sensitization for a given application, but, as seen in Table 2, a value of $P_a = 2 \, C/cm^2$ has been selected as the acceptable maximum for Type 304 stainless steel pipe and tubing in boiling-water nuclear reactors. (The EPR procedure is presently the subject of round robin testing through ASTM.)[39]

TABLE 2

OPTIMUM TEST PARAMETERS FOR CONDUCTING EPR TESTS ON TYPE-304
STAINLESS STEEL[36]

Electrolyte	0.5 M H_2SO_4 + 0.05 M KSCN
Temperature	30 °C
Surface finish	1 μm diamond polish
Reactivation scan rate	3 V/h
Passivation potential	+200 mV for 2 min
Deaeration	N_2 continuous (5 min prior)
Cathodic charging	Only to establish rest potential
Auxiliary electrodes	Pt
Reference electrode	Standard calomel
Polarization system	Any with current integration capability
Data normalization	GBA adjusted
Acceptable (preliminary) material	$P_a^\dagger \leqslant 2$ C/cm^2

$\dagger P_a = Q/GBA$ where $Q =$ charge in coulombs (measured)
$GBA = As|5.095 \times 10^{-3} \exp (0.347x)|$
and, $As =$ sample area (cm^2)
$x =$ ASTM grain size at $\times 100$

Armijo[40,41] has observed IGA in non-sensitized austenitic stainless steels exposed to highly oxidizing nitric acid–dichromate solutions. High-purity alloys were immune[41] but with susceptible commercial alloys increased microhardness was observed at the grain boundaries[40,42] due to solute segregation, particularly of P and Si. Vermilyea *et al.*[43] have also found a similar effect of P and Si on the IGA of a nickel-base alloy (similar to Inconel 600) exposed to nitric acid–dichromate solutions and more recent work[27] has confirmed that solute segregation occurs at grain boundaries of non-sensitized austenitic stainless steels. In highly oxidizing solutions, sulfur at the grain boundaries was found to correlate with IGA sensitivity, but with less strongly oxidizing acid IGA correlated with chromium depletion at the grain boundaries for sensitized material.

5. EROSION CORROSION

Most of the recent work on erosion corrosion has been devoted to copper-base alloys in seawater. Of particular note is the review by Syrett,[44] which covers the available literature on the effects of pH, temperature, dissolved oxygen, entrained gases (bubbles), alloy composition and hardness, pipe diameter and hydrodynamic variables, and water velocity among others. Syrett reports that alloy hardness has very little effect, and that minor alloying elements, Fe, Al, Cr, etc., have considerable effect on erosion corrosion of Cu-base alloys in seawater. Also, it is generally agreed that there is a 'breakaway' velocity above which erosion corrosion attack increases abruptly. These facts suggest that erosion of the surface corrosion product film is controlling rather than erosion of the substrate alloy itself, although calculation[44] of the surface shear stress gives values of only $0.0046 \, N/mm^2$ even at a velocity of 30.5 m/s. Such low stresses may not be sufficient to erode physically any but the most fragile surface films. Thus, Syrett concludes: '... that the so-called erosion of the oxide film may actually be the result of a chemical or electrochemical reaction and hence the result of mass transfer phenomena. Thus, the breakaway velocity may reflect conditions where the rate of film removal (e.g. by dissolution in the sea water) is higher than the rate at which the film can reform by other competing reactions.' It is difficult to accept simple mass transport and dissolution as general mechanisms controlling erosion corrosion in many systems, especially in view of the fact that abrasive slurries clearly cause accelerated attack on iron by a physical erosion process.[45] In every case, whether the abrasive was harder or softer than the iron, the slurry caused accelerated anodic dissolution behavior and it is difficult to avoid the conclusion that a corrosion product film is being stripped from the surface, exposing fresh metal. Syrett[44] also concluded that breakaway velocity is a function of the flow geometry, e.g. pipe diameter, and recommended that studies be conducted under precisely characterized hydrodynamic conditions. One such attempt[46] employed a circling foil apparatus, in which hydro-dynamic flow was well defined, but the velocity was limited to about 3 m/s.

Efird[47] has conducted erosion corrosion tests on copper-base alloys in seawater under well-defined hydrodynamic conditions; a longitudinal section of his apparatus appears in Fig. 8. The cylindrical sample holder ($\sim$ 5 cm outside diameter) had a rectangular flow channel cut through its length with grooves to hold the sheet sample in the center of the channel so that the seawater passed parallel to the surfaces on both sides of the sample. Plastic inserts of the same thickness and width as the test sample were placed in the same grooves upstream and downstream of the latter in the flow establishment and flow maintenance sections, respectively, of the sample holder. In this way, well defined hydrodynamic characteristics were established over the entire length and on both sides of the sample. Surface shear stresses were calculated and are plotted as a function of velocity in Fig. 9. The surface shear stress, a more fundamental

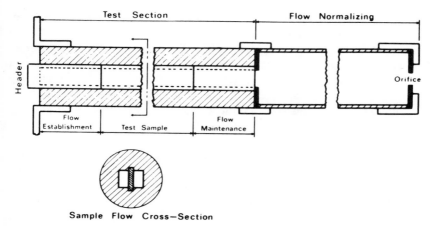

Sample Flow Cross—Section

FIG. 8. Fluid flow materials test apparatus. Cross-section of sample holder for erosion corrosion testing with control of hydrodynamic conditions.[47] By courtesy of the National Association of Corrosion Engineers.

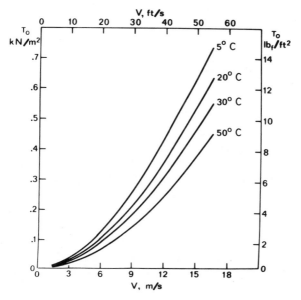

FIG. 9. The surface shear stress, T_0, on test specimens in the sample holder of Fig. 8 as a function of velocity, V, for varying seawater temperature.[47] By courtesy of the National Association of Corrosion Engineers.

parameter, was thus related to the surface velocity, an experimental variable controlled by the upstream pressure at the header and the downstream exit diameter at the orifice. The critical surface shear stress at the breakaway velocity should be the same for any flow geometry. In principle, once the critical shear velocity is known, the breakaway velocity can be calculated for any designed flow geometry. Unfortunately, Efird did not confirm this concept experimentally by obtaining the same critical surface shear stress from two or more different flow geometries.

Erosion corrosion is generally irregular, exhibiting open pits which often coalesce into a wavy directional pattern;[48] an example, of an aluminum surface exposed to rapid flow of hot water, is shown in Fig. 10. Localized erosion corrosion takes the form of teardrop-shaped holes or pits which grow with the large end in the downstream direction. These pits may elongate into grooves, or several may grow side by side having initiated at a scratch or groove in the surface perpendicular to the flow direction; both effects can be seen in Fig. 10. It is believed[49] that erosion corrosion pits initiate at any surface asperity or flow irregularity which can cause localized turbulence, which probably causes a much greater local surface shear stress than that calculated from the macroscopic flow conditions. The localized turbulence erodes the surface product film, causing a local loss of protection and increased corrosion rate which further enlarges the asperity into a cavity (Fig. 11). This, in turn, increases the size and intensity of the local eddy and creates a pit or groove which is self-propagating so that erosion corrosion is found to be maximized at areas of greatest turbulence.[50]

Bianchi et al.,[51] have reported localized 'horseshoe' corrosion at the inlet of copper alloy seawater heat-exchanger tubes. Attack is generally limited to that area within a few diameters of the inlet, again where turbulence is maximized. The authors attribute horseshoe corrosion to localized turbulence at a surface irregularity creating a cavity that stabilizes the local turbulence and propagates the local erosion corrosion. A 'horseshoe' cavity is shown schematically in Fig. 12, the mechanism whereby such grow being similar to that suggested above for aluminum,[49] although the shape of the 'horseshoe' is somewhat different from the 'teardrops' of Fig. 10. Bianchi et al., also suggest that a galvanic effect between the unfilmed pit bottom and the surrounding passivated surface aggravates the attack.

Water chemistry has been found to have a profound effect on the erosion corrosion of aluminum[49] exposed to a small stream (initially

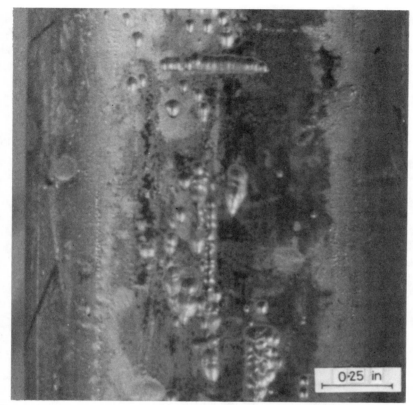

FIG. 10. Erosion corrosion pits and cavities on aluminium exposed to high velocity hot water.[49]

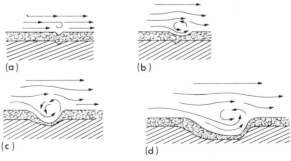

FIG. 11. Turbulent eddy mechanism for growth of erosion corrosion pit.

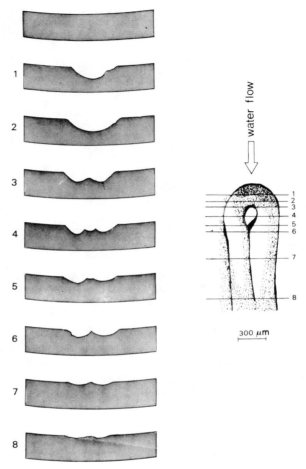

FIG. 12. Actual profile sections of horseshoe corrosion of an aluminum–brass tube exposed to flowing seawater.[51] By courtesy of the National Association of Corrosion Engineers.

0.25 mm × 1.27 mm cross-section) of hot water passing the surface at nominally 27.5 m/s. In distilled water a tightly adherent film was formed that was resistant to any erosion corrosion, but in a well water a dark surface film formed which was swept away in the flow channel leaving a bright unfilmed surface. Aluminum exposed to river water exhibited intermediate behavior, with an adherent film present, but some bright, eroded areas were visible in the flow channel. Table 3 gives data for erosion corrosion in terms of the depth of the eroded channels in the aluminum, showing that there was very little attack in distilled water, but considerable penetration in well water. Typical analyses of the river and

TABLE 3

EROSION CORROSION IN VARIOUS WATERS AFTER 24 h EXPOSURE[8]

Water source and treatment	Average channel depth† (mm)
Distilled water, adjusted to pH 8.3 with NaOH	0.005
River water, adjusted to pH 8.3 with NaOH	0.076
Well water, pH = 8.3, untreated	0.584
Distilled water, 50 ppm $CaCO_3$, nitrogen purge to remove CO_2 and bring pH to 8.3	0.050
Distilled water, 10 ppm SiO_2 (as $Na_2SiO_3.9H_2O$), pH adjusted to 8.3 with H_2SO_4	0.152
Distilled water, 50 ppm $CaCO_3$ and 10 ppm SiO_2 (as $Na_2SiO_3.9H_2O$), nitrogen purge to remove CO_2 and bring pH to 8.3	0.686

†Average of four equally spaced measurements along the length of the channel.

well (tap) waters used in this work are shown in Table 4, indicating that the well water was far less pure, as reflected by higher alkalinity, hardness (mostly due to carbonates), total dissolved solids, chloride, and silicate (measured as SiO_2). The effect of each of these factors on erosion corrosion was measured by additions to distilled water but none showed any major effect, although silicate additions did produce a brightly etched flow channel without significant attack in the flow channel (Table 3). Electron microprobe analysis showed that Si was the only substance present in the corrosion product films on the specimens exposed to well water and river water and not present in the films resulting from distilled

TABLE 4

ANALYSES OF HANFORD TAP (WELL) WATER AND
COLUMBIA RIVER WATER, ppm

	Tap Water	River Water
Methyl Orange Alkalinity	182	58
Hardness	162.5	69
Total Solids	266	94
SO_4^{-2}	26	11
Cl	8.2	0.5
SiO_2	34	5.3
Cu	0.010	0.006
Mg	7.2	4.2
Ca	31.2	21

†R. L. Dillon and R. S. Hope, *Erosion-corrosion of aluminum alloys*, HW-74359 REV (1963) April.

water exposure. It was then found that a combination of bicarbonate and silicate in distilled water would simulate the heavy erosion corrosion attack previously observed in well water (Table 3), suggesting that there was a synergism between bicarbonate and silicate to form a friable film which was easily eroded from the surface.

The bulk of the experimental data seem to support a mechanism whereby a surface film is eroded by impingement of the otherwise film-forming corrosive environment or by abrasive particles carried by the environment. While mass transport may play a role in corrosion of the unfilmed surface, it is not the controlling process, in general. It is also notable that others[47,51] have calculated higher surface shear stresses than the low values given by Syrett.[44] However, increased mass transport of dissolved reactants would tend to stabilize rather than dissolve most surface film and there seems to be little evidence in the literature to support a mass transfer mechanism. Nevertheless, there is a lack of unambiguous data on which to base any mechanism and none of the opinions is incontestable. Definitive experiments to determine the general mechanisms of erosion corrosion are overdue.

6. PITTING AND CREVICE CORROSION

6.1 General

Pitting and crevice corrosion are discussed together because their mechanisms of growth and propagation are generally considered to be the same.[3] Many forms of localized corrosion apparently share the common feature of an occluded anode.[52] A crevice becomes a preferred site for an occluded anode and localized corrosion because the crevice interior becomes starved of dissolved oxygen, and corrosion product cations (e.g. Fe^{2+}, Cr^{3+}, Ni^{2+} in stainless steel) accumulate due to restricted mass transport to the bulk solution,[3,53] with anions, particularly Cl^-, migrating into the crevice to neutralize the positive cations. The resultant salts hydrolyze to form weak bases and hydrochloric acid, so that the contents of the crevice become acidic, which further accelerates corrosion in an autocatalytic manner. The same processes occur in a pit, and crevice corrosion is often thought of as a special case of pitting with a ready-made initiation site. The accelerated corrosion within the occluded anode liberates electrons, ($M = M^{+n} + ne^-$), which must be consumed by reduction of some dissolved species at the surrounding cathode surfaces, e.g. reduction of dissolved oxygen ($O_2 + 2H_2O + 4e^- = 4OH^-$).

Taylor[53] has recently shown that sulphates will concentrate in crevices and locally decrease the pH, particularly at elevated temperatures. This is of concern in causing IGA and cracking in crevices in certain alloys. A 'transient' type of pitting has been observed[54] with iron in sulphate solution, but the localized attack was not propagated for an extended time, while sulfates have also been shown to increase anodic dissolution when concentrated at the surface of austenitic stainless steel.[55]

The remainder of this discussion will be concerned with the differences between initiation, mainly of pitting, and propagation, in which there is much in common between pitting and crevice corrosion, with a treatment of critical pitting, E_{pit}, and protection, E_{prot}, potentials. Emphasis will be placed on developments over the last decade during which period there have been excellent reviews by Szklarska-Smialowska[56] and France[57] with references to previous work on pitting and crevice corrosion, respectively.

6.2 Initiation

Pitting initiates at a critical potential, E_{pit}, which is used as a measure of an alloy's resistance to pitting corrosion since the more noble E_{pit}, the more resistant is the alloy.[56] Wilde and Williams[58] showed good correlation between E_{pit} and the resistance of a series of Fe–Cr alloys to pitting in long-term exposures to seawater. Pitting initiated in any alloy when E_{pit} was exceeded although the time and potential above E_{pit} before initiation were not always reproducible. At initiation the corrosion potential reached a maximum and thereafter continuously drifted in the active direction during propagation (Fig. 13), the latter continuing even at potentials active to E_{pit} in most cases (Table 5). Wilde and Williams[58] showed further that pitting resistance is proportional to the chromium content of the alloy. Thus, it is not surprising that sensitized stainless steels, with chromium-depleted grain boundary areas, are less resistant to pitting[59,60] as well as to IGA.

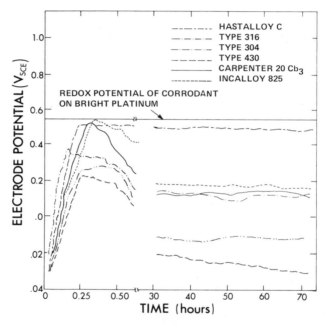

FIG. 13. Potential v. time behavior of alloys exposed to acidified $FeCl_3$ solution.[68] By courtesy of The Electrochemical Society.

The determination of E_{pit} can be rather uncertain because of the difficulty in initiating pitting through a passive film. Stolica[61] observed that surfaces with non-steady state passive films were less resistant to pitting than those which had been prepassivated and stabilized for a long time period before exposure to chloride, with poor reproducibility of E_{pit}. Leckie and Uhlig[62] overcame this difficulty by potentiostating the specimen at successively more noble potentials near E_{pit}, the latter corresponding to the potential where the potentiostatic current

TABLE 5
EXPERIMENTAL DATA ON ALLOYS USED IN CHEMICAL PITTING TESTS IN AQUEOUS
FERRIC CHLORIDE[58]

Alloy designation	E_{pit} (in 1 M $NaCl + N_2$) (V_{SCE})	Corrosion potential during test (V_{SCE})	
		Maximum	Minimum
Type 430 stainless steel	−0.130	0.230	−0.310
Type 304 stainless steel	−0.020	0.280	−0.140
Type 316 stainless steel	−0.100	0.385	−0.090
Carpenter 20 Cb	−0.500	0.520	−0.120
Incoloy 825	−0.525	0.530	−0.180
Hastelloy C	>0.900	0.530	−0.530

increased, marking pit initiation, in less than 24 h. Such a time-consuming procedure is less than satisfactory, and Pessal and Liu[63] proposed the 'scratch method' for determining E_{pit}. An abrasive stylus was inserted into the cell and the specimen surface was scratched at successively higher potentiostatic potentials, E_{pit} being equated to the potential above which the surface would not repassivate after scratching (Fig. 14).[64] Below E_{pit} the surface repassivated after the current spike caused by scratching. For the ferritic[65] and austenitic[66] stainless steels the scratch method gave the least noble and most reproducible values for E_{pit}.

France and Greene[67] found differences between E_{pit} determined chemically with dissolved oxidizers and electrochemically with a potentiostat. They attributed the differences to electrostatically assisted migration of Cl^- causing an artificially elevated local Cl^- concentration which lowered E_{pit} when determined potentiostatically. However, it was later found[68] that E_{pit} is made more noble by purging the solution with oxygen and when dissolved oxidizer was used which did not interact with the passive film good correlation was found between chemically and electrochemically determined E_{pit}. Clearly the latter cannot be used as an absolute indicator of potential ranges in which pitting will or will not occur, because of possible variations due to dissolved gases. It has been confirmed subsequently[69] that there is a local surface concentration of Cl^- as potential increases and approaches E_{pit}, no matter whether the changes are caused chemically or potentiostatically. Ultrasonic vibration apparently disperses this surface chloride because E_{pit} is made more noble for stainless steel[70] and aluminum,[71] but the chloride must be fairly strongly bound to the surface because simple surface rotation had

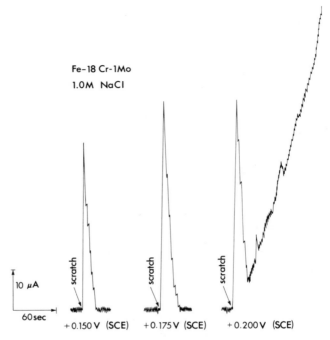

FIG. 14. Current transients after surface scratching below and above (0.200 V) the critical pitting potential.[64] By courtesy of The Electrochemical Society.

no effect on E_{pit} for stainless steel.[72] Janic-Czakor *et al.*[73] have in fact found evidence of relatively thick chloride salt 'islands' on the surface of pure iron at potentials even below E_{pit}. These islands are presumably the nucleation sites for pitting and ultrasonic vibration could break up the islands while rotation (solution flow) could not. The microenvironment beneath an island will approach that of a fully developed pit and initiation can proceed much the same as does propagation. Augustynski *et al.*[74] showed that only the anions which form highly soluble salts cause breakdown of passivity so that salts deposited on a surface must have a high associated local surface electrolyte concentration, which leads to the acidity necessary to initiate pitting. Strehblow and Ives[75] have concluded that a salt film would be necessary during initiation and early growth of a pit, because any ohmic losses in a small pit are not sufficient for stable growth (see subsequent discussion, Section 6.3). It is not known exactly how these salt islands interact with the passive film although Auger electron spectroscopic (AES) analysis of passive films[76,77] has shown that, generally, chloride is bound only to the exterior of the film with little or no penetration. However, a localized penetration prior to formation of an island is still possible.

An exchange of Cl^- with other anions composing the passive film[73] would explain the results from a rotating ring-disc by Heusler and Fischer[78] who found, for a constant anodic potential up to initiation of pitting, that Fe^{3+} was produced without any coincident change in the passive anodic current after Cl^- had been injected into the electrolyte. A possible exchange reaction might be

$$FeOOH + Cl^- = FeOCl + OH^-$$

where FeOOH represents the hydrated passive film with iron in the ferric oxidation state. The ferric oxychloride, FeOCl, would have a considerably higher solubility than the FeOOH and would dissociate as follows:

$$FeOCl + H_2O = Fe^{3+} + Cl^- + 2OH^-$$

producing the Fe^{3+} detected in Heusler's experiments in which the localized formation of HCl would contribute to the initiation of the localized corrosion.

Szklarska-Smialowska et al.[79,80] and others[81] have shown that pits generally nucleate at mixed $(MnFe)S_x$ sulfide inclusions, but S additions alone, without any concomitant Mn, had relatively little effect on pitting resistance.[79,82] However, the most susceptible sites were those in which the sulfides were closely associated with oxides of aluminum or chromium.[79,83] Manning et al.[84] have shown that sulfide exposure as affected by edges and surface grinding will decrease resistance to pitting corrosion. Similarly, Mazza et al.[66] propose that cold work fragments sulfide inclusions, exposes more sulfide area at the surface and lowers E_{pit}, but while some workers report an adverse effect of cold work on pitting,[65,85] others have found no effect.[86,87] The remarkable resistance of the rapidly solidified amorphous alloys to pitting may be due to their homogeneity and the absence of non-metallic inclusions.[88,89] Retained ferrite in duplex stainless steels initiates pitting at the austenite–ferrite boundaries, lowering E_{pit}.[83] On the other hand, strain-induced martensite in the so-called TRIP steels increases pitting resistance.[90]

The mechanism of nucleation at inclusions is uncertain. Preferential dissolution of the sulfide inclusion may create a crevice at the sulfide/metal interface[78] which serves as the initiation site. Ecklund[91] has found that crevice corrosion initiates at sulfide inclusions and he postulates that sulfides are electrochemically active and are preferentially dissolved by anodic polarization when coupled to the matrix metal. The recent work of Gainer and Wallwork,[92,93] using a scanning microprobe potentiometer to map surface potentials on mild steel, showed sulfide inclusions to be active relative to the surrounding steel. Inclusions consisting of $(Mn,Cr)S$, $MnS + Cr_2S_3$ or $MnS + Al_2O_3$ were more active electrochemically than the commoner $(Mn, Fe)S$ inclusions in steel. The oxides alone were noble with respect to the matrix but were much more active than the matrix when combined with sulfides, while additions of Cu and Ni modified the matrix potential to improve pitting resistance. Thermodynamic calculations[94] support a sulfide dissolution mechanism of pit initiation over one involving oxides.

It is assumed that salt islands form randomly on the surface, independent of any surface inclusions which may be present. Janic-Czakor et al.[73] observed salt islands on pure iron surfaces without any detectable non-metallic surface inclusions. However, when surface inclusions are present, it may be inferred that pits initiate at the inclusions under the islands. Sulfides increase the critical current density for passivation in the active corrosion region of stainless steels in the acid solutions which are to be expected in the microenvironment under each salt island.[82,95] Thus, sulfide inclusions accelerate the acid dissolution process and facilitate initiation of pits under the salt islands.

Galvele[96,97] has proposed that there is a product, Xi, of pit depth (X) and current density at the pit (i), which is critical to initiate and stabilize a pit. The generation of a salt film island on the surface creates a concentrated acid chloride microenvironment beneath the island and a concentration cell with the surrounding passive surface. The active sulfide inclusions under an island are rapidly attacked, creating the necessary X to initiate and stabilize pits. Only one such pit can survive in a given area, becuase the cathodic polarization from the strongest pit soon passivates all surrounding nearby pit nuclei.

6.3 Propagation

Just as E_{pit} is associated with the initiation of pitting corrosion, the 'protection potential', E_{prot}, is associated with growth and propagation of pitting corrosion. Pourbaix[98] defined E_{prot} as the potential below which existing pits cannot propagate, so that there is a range of potentials, $E_{pit} - E_{prot}$, in which pits will not initiate but existing pits will propagate. E_{prot}, along with E_{pit}, is usually determined by a cyclic polarization procedure, in which the passive alloy is polarized potentiodynamically above E_{pit} to a predetermined current, at which time the potential scan is reversed. As is apparent from Fig. 15,[58] the reverse potential scan produces considerable hysteresis, and E_{prot} occurs at the potential where current is reduced to the original passive value. Pourbaix[98,99] has suggested that E_{prot} is near the actual potential of surfaces within the pit, which must vary with acidity and chloride concentration. Thus, E_{prot} decreases with the time of the preceding pit propagation[100-102] and consequently E_{prot} is not a unique property of an alloy, depending on experimental procedure, and must be used with extreme caution in characterizing alloys, as is sometimes done.[103]

Wilde and Williams[101,102] have shown that the resistance of stainless alloys to marine corrosion cannot be judged only by E_{pit}, because it is a measure of pit initiation only. In practice crevice corrosion is the major mode of attack because it is more easily initiated in the sheltered crevice area and a meaningful test must measure resistance to propagation in the acid chloride solutions generated within crevices. The 'difference' potential $E_{prot} - E_{pit}$, as determined by cyclic polarization (Fig. 15), correlates with resistance to long-term marine attack by crevice corrosion.[101,102] Both E_{prot} and E_{pit} are highly dependent on experimental procedure and the difference potential is subject to the same difficulties, so that strict duplication of experimental procedure is required before $E_{pit} - E_{prot}$ can be compared for different alloys.

For some alloys there was no breakdown potential, E_{pit}, and the difference potential could not be measured.[101] In such cases, cyclic polarization was conducted on a specimen containing an artificial crevice and the resulting value of $E_{pit} - E_{prot}$ was found to be dependent on the size and geometry of the crevice relative to the rest of the specimen. Thus, any electrochemical characterization of crevice corrosion[104-106] applies strictly only to the identical specimen geometry and procedure. Thus, Kain[107] found that alloys were ranked in a different order of resistance if the crevice geometry in laboratory tests did not duplicate that in marine exposure tests.

Syrett[108] has proposed a pitting propagation rate (PPR) test at potentials between E_{pit} and E_{prot}, the procedure being as follows with reference to Fig. 16. The specimen is polarized to a potential E_{test} (0.25 V in Fig. 16) between E_{pit} and E_{prot} and the steady state passive current,

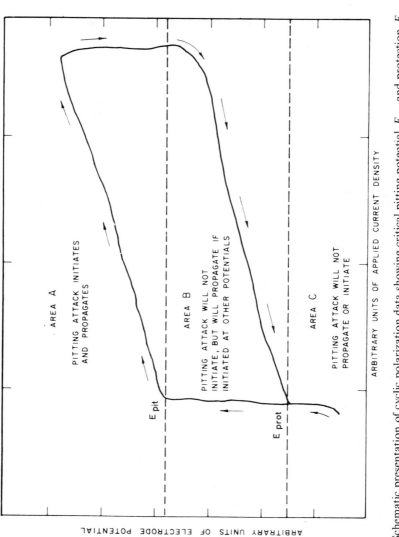

FIG. 15. Schematic presentation of cyclic polarization data showing critical pitting potential, E_{pit}, and protection, E_{prot}.[58] By courtesy of The Electrochemical Society.

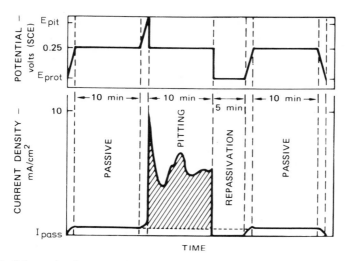

FIG. 16. Schematic of potential–time and current–time in the PPR procedure.[108] By courtesy of the National Association of Corrosion Engineers.

I_{pass}, is recorded after 10 min. The potential is then scanned to a value above E_{pit} where pits begin to initiate and when the current density reaches a nominal 10 mA/cm², the potential is reduced in a single step back to E_{test}, following which the current is recorded during a 10 min hold time. The specimen is then allowed to return to the free corrosion potential for 5 min, deactivating the pits, after which the potential is returned to E_{test} for 10 min to check that I_{pass} is still the same as at the beginning of the test. The procedure is repeated for a number of E_{test} values between E_{pit} and E_{prot} and at each E_{test} a current I_{test} is calculated by dividing the charge passed at E_{test} (from graphical integration) by the 10 min hold time. A pitting current, I_{pit}, is then obtained from $I_{pit} = I_{test} - I_{pass}$ and the PPR is finally calculated as $i_{pit} = I_{pit}/A_{pit}$, where A_{pit} is the microscopically determined pit area.

Pitting progagation rate (PPR) is plotted as a function of potential for a TRIP steel with varying amount of cold work in Fig. 17, from which it is apparent that the potential of zero pit growth, E_{ZPG}, is considerably more noble than E_{prot} for both cold-worked conditions indicating an expanded range of usefulness over that suggested by E_{prot} alone. However, it may be that the E_{ZPG} approaches E_{prot} as the degree of occlusion (amount of pit propagation) increases. The difference between E_{prot} and E_{ZPG} is affected by alloy composition[109] and may reflect the longer

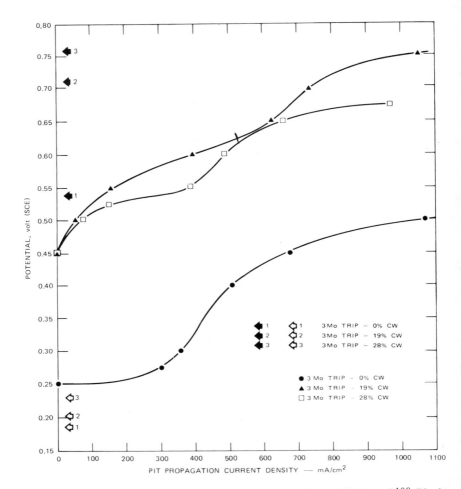

FIG. 17. Effect of cold work on the PPR curve for 3Mo TRIP steel.[108] Black arrows, mean E_{pit} values; white arrows, mean E_{prot} values; determined by the cyclic polarization method. By courtesy of the National Association of Corrosion Engineers.

pit propagation time in the cyclic polarization procedure. It may be possible to make E_{prot} equivalent to E_{ZPG} by increasing the sweep rate in the cyclic polarization test and it would appear that more work is needed before the utility of the PPR test can be judged.

After the initiation of pitting[58] or crevice corrosion[104,110] in a chemical test (no potentiostatic control) it is found that the overall potential of the specimen falls to progressively more active values with increasing propagation time (Fig. 13). This behavior has been attributed to progressively higher acidity and chloride concentration during the growth of localized corrosion.[111] In a freely corroding test, it was found that the pit or crevice anode is polarized only a few millivolts, while the passive external cathode is polarized over 150 mV to the overall corrosion potential, E_{corr}, of the specimen (Fig. 18). This is consistent with the previous conclusion[98,99] that the protection potential, E_{prot}, is very near the potential within the pits. Increased acidity and chloride bring the anode to potentials several hundred millivolts more noble than may be found in more dilute solutions, so that the same potential is active for the acid chloride solution within a pit or crevice, but is in the passive range for the exterior surfaces exposed to the dilute bulk solution.

The necessity of concentrated solutions within the occluded anode was

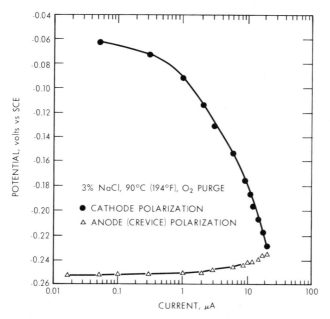

FIG. 18. Simultaneous anodic polarization of the occluded anode and cathodic polarization of the passive surface during localized corrosion of Type 304 stainless steel.[111] By courtesy of Pergamon Press Inc.

emphasized by the work of Rozenfeld and Danilov[112] who showed that when the covering cap of corrosion products over a pit was damaged, permitting dilution of the interior, the pit deactivated. In model pit studies, Suzuki et al.[113] demonstrated that the total cation (mainly Fe^{2+}, Cr^{3+}, Ni^{2+}) concentration exceeds 6 N and the corresponding pH is < 1.0 in an occluded anode. Mankowski and Szklarska-Smialowska[114] collecting and analyzing the pit anolyte, found chloride concentrations up to 12 N which decreased to 2 N as pitting propagated. Concentrated solutions of iron, chromium and nickel chlorides, simulating those found in the pit anolyte, hydrolyze to very low pH,[111,114] as shown in Table 6.[111]

TABLE 6
pH OF CONCENTRATED CHLORIDE SALT
SOLUTIONS AT ROOM TEMPERATURE[111]

Salt	pH at indicated salt concentrations		
	1 N	3 N	Saturated
$NiCl_2$	3.0	−2.7	−2.7
$FeCl_2$	2.1	−0.8	−0.2
$CrCl_3$	1.1	−0.3	−1.4

Thus, in summary, E_{prot} is attributed to the formation of an occluded anode during initiation of pitting and when the potentiostatic current which initiated pitting is reversed, the polarization curve follows a path characteristic of the occluded anode, variable hysteresis being observed depending on the degree of developed occlusion and the resistance of the alloy to the acidic chloride solutions of the pit interior. At zero anodic current the potential is near that of the open circuit potential, E_{anode}, within the pit or crevice. When localized crevice or pitting is present with dissolved oxidizers, i.e. no potentiostatic control, the same occluded anode develops with a concentrated acid chloride solution in the pit or crevice. The occluded pit/crevice anode is polarized several millivolts above E_{anode} by galvanic coupling to the surrounding cathode where the dissolved oxidizers are reduced. When a potentiostatic polarization above E_{pit} is removed, previously initiated pits become inactive because the potential falls below E_{anode} in the absence of dissolved oxidizers. This agrees with the observation of Kitamura and Suzuki[115] that pit growth occurs only at potentials more noble than the open circuit anode potential of the pit interior.

It has been difficult previously to understand how the active state can be maintained within a pit or crevice while the outer passive surface retains a noble passive potential. A large ohmic potential drop within the pit has been postulated[116] to account for this, but Fig. 18 shows that nearly all of the polarization between the passive exterior and the interior occurs at the exterior passive surface. This verifies the previous work of Rozenfeld and Danilov[112] who showed strong cathodic polarization of adjacent surfaces *outside* of a pit. The potential distribution curve in the vicinity of an active pit (Fig. 19) shows further that the

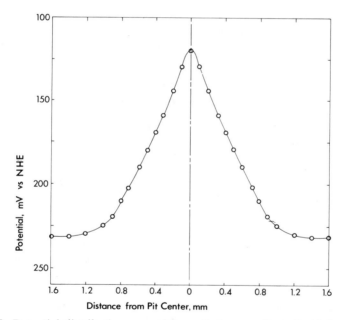

FIG. 19. Potential distribution around an actively corroding pit. 18-8 stainless steel in 0.04 M $FeNH_4$ (SO_4). $12H_2O + 0.56$ M NH_4Cl. By courtesy of Pergamon Press Inc.

potential returns to the unpolarized passive potential at points relatively remote from the pit anode. If there were a significant ohmic drop in the pit itself, it would be evident in Fig. 18 as marked anodic polarization. Strehblow and Ives[117] found the ohmic drop assumption untenable for small, newly initiated pits and postulated a salt film as responsible for stabilizing the early growth of pits, much as was discussed previously.

Other investigators[118] have proposed that the difference $E_{pit} - E_{prot}$ is due to an ohmic drop through the pit but again, the low overvoltage at the occluded anode (Fig. 18) is not consistent with such an idea, nor can the hysteresis in a cyclic polarization curve (Fig. 15) be explained by ohmic drop effects. Thus, at the conclusion of a cyclic polarization test there is almost no current, no ohmic drop, and E_{pit} should be very nearly the same as E_{prot}, which is not the case.

Some investigators[119] have defined an E'_{prot} below which a scratched surface will repassivate, which is similar to the E_{pit} determination by scratching proposed by Pessal and Liu.[63] However, occluded anode potentials may be expected to reach more active values than would be measured on a scratched surface exposed to the dilute bulk solution[108] so that $E'_{prot} \neq E_{prot}$. Furthermore, the E_{pit} determined by scratching was found to bear no relation to crevice corrosion resistance with fully developed occluded cells present.[120] Thus, alloys high in Cr and Mo had less resistance to crevice corrosion although E_{pit} by scratching was equal to or higher than that of other stainless steels, probably because the chloride salts of Cr and Mo hydrolyze to even lower values of pH in an occluded cell than either Fe or Ni.[111,113,114] Although E_{pit} by scratching may give a good indication of resistance to pit initiation due to Cr and Mo additions, in relation to crevice corrosion resistance it is poor due to the acid corrosion products of these two elements generated in the occluded crevice cell.

As crevice or pitting corrosion propagates, the specimen potential falls to more active values,[104] corresponding to ever increasing acidity and chloride in the occluded crevice cell. The minimum potential will be reached when the salts of all elements of the alloy reach saturation in the pit or crevice. Neglecting the small anodic polarization at the crevice anode, this minimum potential should correspond to the crevice protection potential, discussed by Verink et al.,[121] below which crevice corrosion can neither initiate nor propagate. A crevice of specific geometry may not reach the minimum potential, due to migration of corrosion products out of the occluded anode preventing saturation, and, in principle, E_{prot} for pitting should also reach this same minimum in the limit, but geometric factors may again prevent saturation in the pit. Nevertheless, the most conservative value of E_{prot} for both pitting and crevice corrosion would be the corrosion potential in the saturated salts of the alloy elements.

Most of the discussion above pertains to ferritic and austenitic stainless steels, but the same general theory also seems applicable to aluminum, although the two systems may differ somewhat in detail.[122] For example, aluminum alloys display well-defined hysteresis under cyclic polarization with easily measured values of E_{pit} and E_{prot}. Furthermore, de Wexler and Galvele[123] have postulated that localized corrosion of aluminum cannot propagate unless the potential is above the corrosion potential of aluminum in saturated $AlCl_3$.

The conclusions of Galvele,[96,97] based mostly on non-ferrous alloys, are essentially similar to those indicated above for ferrous alloys. Galvele proposes that pitting potential, E_p, is given by

$$E_p = E_{corr,A} + \eta + \phi + E_{inh}$$

$E_{corr,A}$ is the open circuit corrosion potential of the occluded anode in the pit or crevice; η is the anodic activation overvoltage at the occluded anode caused by coupling to the external passive surfaces or imposed externally with a potentiostat; ϕ is the ohmic potential drop caused by a pitting current, I_{pit}, passing through the resistance within the pit or crevice; and E_{inh} is the overvoltage caused by inhibitors, buffers or passivators in the solution. A pitting potential determined by potentiodynamic scanning always contains an overvoltage, $E_{pit} - E_{prot}$, necessary to initiate pitting and even surface scratching does not reduce E_{pit} to a value equal to E_{prot}.[120] Thus, Galvele's pitting potential, E_p, is more properly designated as the protection potential, E_{prot}, as usually used in the literature.

In chloride solutions with no buffer or inhibitor ions present, ϕ and E_{inh} are very low,[111] so that Galvele's equation can be simplified to

$$E_{prot} = E_{corr,A} + \eta$$

For activation-controlled kinetics the anodic overvoltage, η, follows the Tafel relationship

$$\eta = \beta_a \log i_a/i_o$$

where β_a is the anodic Tafel constant, i_a is the anodic current density or anodic dissolution rate, and i_o is the exchange current density for the anodic reaction. For most anodic dissolution reactions β_a is quite low, of the order 30–50 mV/decade, as confirmed for the occluded anode reaction by the very low slope in Fig. 18. Thus, if $\eta = 30$ mV, a cathodic potential 30 mV active to E_{prot} reduces i_a by a factor of 10 and 60 mV by a factor of 100.[111,124] Considering that corrosion products are constantly migrating out of the occluded anode area, the occluded anode must be constantly fed by the anodic dissolution reaction to maintain the necessary acid chloride conditions within the pit or crevice, so that even slight reductions in the anodic dissolution rate at potentials below E_{prot} have a profound effect on the activity of the occluded anode.

6.4 Summary

The electrochemical conditions in a growing pit or crevice can perhaps be best summarized by the schematic polarization diagram of Fig. 20, in which potential current and time are given on arbitrary scales. The metal, M, is presumed to be passivated by reduction of some dissolved species Z^{2+} (an oxidizer), which might be dissolved oxygen, but an arbitrary species Z^{2+} is selected for generality. The corrosion potential and corrosion current are determined by the intersection of the anodic dissolution process and the cathodic reduction process, with the latter presumed to reach a limiting current $i_{Lim,Z^{2+}/Z^+}$ for diffusion of Z^{2+} to the electrode surface. As the passive film grows at increasing times, t_1, t_2 and t_3, the anodic dissolution current and passive corrosion current decrease with time while the corrosion potential becomes more noble (points 1, 2 and 3). At 3, the potential exceeds E_{pit} and an occluded cell initiates, but the amount by which the potential exceeds E_{pit} before pitting initiates is not particularly reproducible. The value of E_{pit} would be somewhat more active in a crevice as compared to a free surface and would vary depending on crevice geometry, temperature, etc. Mechanical damage of the passive surface film will also cause a more active value for E_{pit}.[63] As the potential approaches E_{pit} the chloride at the surface increases,[69] forming islands of MCl salt.[73] The interior of a crevice if present is starved of oxidizer, Z^{2+}, and chloride ions migrate into the crevice, accelerating the initiation process in such an area. Localized corrosion initiates at non-metallic inclusions (manganese–iron sulfides with and without chromium and aluminium oxides for ferrous alloys)[79,93] in the acid chloride microenvironment beneath the salt islands. Where a pit has just become stable the anodic process in the 'occluded anode' so formed is represented by a relatively flat (low overvoltage) anodic curve beginning at the free corrosion potential, $E_{corr,A,4}$ (initiation), and terminating at point 4 in Fig. 20. Liberation of the metal cations of the alloy salts further decreases the pH. The decreased pH and increased chloride concentration in the pit or crevice ('increasing occlusion') pushes $E_{corr,A}$ to progressively more active values at times t_4 to t_8 and the corrosion potential of the specimen drifts in an active direction with time after initiation. The plot of corrosion potential v. time (Fig. 20) qualitatively resembles the experimental data of Wilde and Williams[58] for stainless

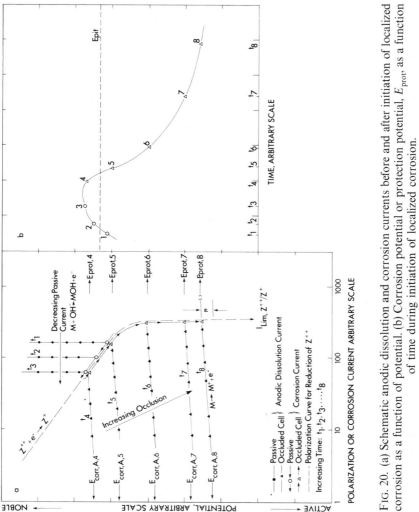

FIG. 20. (a) Schematic anodic dissolution and corrosion currents before and after initiation of localized corrosion as a function of potential. (b) Corrosion potential or protection potential, E_{prot}, as a function of time during initiation of localized corrosion.

steel in ferric chloride solutions (Fig. 13), and reflects the longer time required to reach high levels of occlusion as outward migration of corrosion products becomes significant. The corrosion current after initiation of localized corrosion may be expected to climb to a maximum value corresponding to the limiting diffusion rate of the oxidizer, Z^{2+}, to the cathode surfaces, in agreement with experimental results[104,111] for crevice corrosion of austenitic stainless steel.

The measured specimen corrosion potential is the same as that in the occluded anode (neglecting any small ohmic losses) and may be designated as the protection potential, E_{prot}. As time and degree of occlusion increases, E_{prot} decreases, as does the corrosion potential. Any potential active to E_{prot} applies cathodic protection to the occluded anode and immediately suppresses the localized corrosion therein.[111] The minimum value of corrosion potential (and E_{prot}) corresponds to the approach of alloy element salts to saturation in the occluded anode at t_g. The open circuit anode corrosion potential for saturation in the occluded anode is given by $E_{corr,A,8}$ and differs from the protection potential $E_{prot,8}$ only by the anodic polarization η at the occluded anode. It follows that localized corrosion (whether crevice or pitting) cannot propagate below $E_{corr,A}$ at saturation. Therefore, the most conservative value of the protection potential, E_{prot}, is given by the corrosion potential of the alloy in a saturated solution of the chloride of the major alloy element.

APPENDIX

A.1 Denting Corrosion

The problem of denting corrosion in PWR steam generators was briefly described early in this review[1,125,126] (Fig. 1). Primary cooling water recirculates from the reactor core through Inconel tubing in the steam generators. The secondary water is heated to 290 °C at 1000 psi by heat transfer from the Inconel tubing in the steam generators. The secondary water is then boiled to steam which drives the power generating turbines. Following the turbines the water is condensed, purified in resin beds (in some designs), and returned to the steam generators. The water picks up various impurities from the condensers, the resin beds, make-up water system and the various additives for control of dissolved oxygen, pH and general water chemistry. Some of these impurities and their effects are summarized in Table 7.

The localized corrosion is due to an occluded cell in the crevice between the carbon steel tube support and the Inconel heat exchanger

TABLE 7

COMMON SOLUTES IN STEAM GENERATOR WATER AND THEIR EFFECTS ON LOCALIZED CORROSION AND CREVICE ATTACK

Component	Source	Effect
NH_4^+, pH (AVT when combined with N_2H_4)	Deliberate addition to maintain slightly basic pH. By product of deoxygenation by N_2H_4	Does not concentrate OH^- in crevices because it decomposes to volatile NH_3 and escapes
Cl^-	In-leakage from upstream seawater cooled condensers	Concentration in crevices causes acid hydrolysis, accelerated corrosion of tube supports, and denting of tubes
Dissolved oxygen, DO	Upstream secondary water conditioning and handling	Electrochemical reduction probably required in a galvanic couple to promote and sustain crevice attack
Na^+	In-leakage from condensers, deliberate addition as phosphate salts or leached from condensate polishers.	Concentration in crevices brings concomitant OH^-, caustic concentration, and caustic cracking
Cu^{2+}	Corrosion of brass and admiralty condensers, in-leakage of condenser cooling water.	Reduced to Cu metal in steam generators and collects in sludge on tube sheets. May act as oxidant in absence of DO to promote crevice attack
Boric acid	Deliberate addition	Neutralize crevice acid, plug pores in crevice magnetite.
Fe^{2+}, Fe^{3+}, Ni^{2+}	Corrosion of SG tubing, tube supports and tubesheets	Fe^{3+} may act as oxidizer to support crevice attack in absence of DO. Iron and nickel (and copper) have accelerated denting in lab boiler studies.
Hydrazine (N_2H_4)	Deliberate addition to remove dissolved oxygen	Has not eliminated denting. Reaction with DO apparently rather slow at low temperatures in feed-water train.
Sulfate	In-leakage from upstream condenser cooling water (e.g. from Great Lakes or cooling towers)	Concentration in crevices may cause either caustic cracking and intergranular attack or acid dissolution (wastage).
Phosphate	Deliberate addition as buffer to control pH	Reaction with magnetite deposits may cause pitting, wastage, intergranular attack, or caustic cracking depending on $Na:PO_4$ ratio.

tubes. Chloride impurities concentrate in the crevice by migration to neutralize the cation corrosion products as described previously for usual crevice corrosion. Heat transfer is relatively low in the crevice and salt concentrations are further increased by localized boiling and evaporation. Crevice effects are further aggravated by galvanic interaction between the tubes and tube support. The carbon steel acts as the anode in the galvanic couple, and bulky ferrous oxide corrosion products as described by Potter and Mann,[127] are generated in the crevice. Accumulation of these oxide corrosion products develops stresses suffcient to deform mechanically the tubes and crack the tube supports.

Sodium phosphate was formerly added to secondary water as a buffer to control pH. Slow rates of denting were first observed in PWRs using phosphate water treatment, but elimination of phosphate has seriously aggravated denting corrosion. Sodium phosphate water treatment has caused a wide range of corrosion problems with Inconel 600 tube material depending on the ratio of Na to PO_4 resulting from reactions with magnetite deposits. Attack varies from pitting, to uniform rapid dissolution, to intergranular attack, to caustic cracking as the $Na:PO_4$ ratio progresses from 1.6 to 3.0.[126] Sodium itself may also cause a problem. Accumulation of Na^+ in boiling crevices can result in high pH and danger of caustic stress cracking.[128] Consequently, an all-volatile treatment (AVT) has been initiated in most steam generators tubed with Inconel 600. The AVT consists of hydrazine, N_2H_4, to eliminate dissolved oxygen (DO) and ammonia to control pH. Excessive heating decomposes NH_4^+ and N_2H_4 to ammonia gas, NH_3, which is volatile and consequently cannot accumulate within crevices. Dissolved oxygen must be reduced to very low levels to control corrosion generally and is believed to contribute to denting by providing an oxidizer for the Inconel cathode surfaces outside the tube support crevice. Various cationic impurities (Fe^{3+}, Ni^{2+}, Cu^{2+}, Pb^{2+}) may also act as oxidizers to accelerate denting in the absence of DO.

Boric acid and calcium hydroxide additions are under consideration to neutralize crevice acid and plug the pores in the crevice oxide.[125] Periodic low temperature soaks or flushing procedures are also possible to moderate crevice temperature, leach concentrates out of crevices and thereby slow denting attack.

Denting corrosion is of considerable practical importance in the many PWRs currently in service and under design. Thus, the problem is presently the subject of intensive investigation.

REFERENCES

1. W. H. Layman and L. J. Martell, 'Status of Steam Generators,' Electric Power Research Institute, Palo Alto, CA, presented at the 1979 American Power Conference.
2. H. H. Uhlig, Corrosion and corrosion control, 2nd edn, Chapters 3 and 4, Wiley, New York (1971).
3. M. G. Fontana and N. D. Greene, Corrosion Engineering, 2nd edn, Chapters 9 and 10, McGraw-Hill, New York (1978).
4. F. Mansfeld, D. H. Hestenberg and J. V. Kenkel, Galvanic corrosion of Al alloys: I. Effect of dissimilar metal, Corrosion, 30, 343 (1974).
5. M. C. Reboul, Galvanic corrosion of aluminum, Corrosion, 35, 423 (1979).
6. M. S. Walker, The galvanic corrosion behavior of dissimilar metal couples used in automotive applications, Materials Performance, 18, 9 (1979) April.
7. D. A. Jones, Monitoring galvanic short-circuit current, Electrochemical Technology, 6, 241 (1968).
8. D. A. Jones, unpublished results.
9. F. Mansfeld, Area relationships in galvanic corrosion, Corrosion, 27, 436 (1971).
10. J. Devay, B. Lengyel jun and L. Meszaros, Acta Chimica Acad. Sci. Hungaricae, 62, 157 (1969).
11. G. Lauer and F. Mansfeld, Measurement of galvanic corrosion current at zero external impedance, Corrosion, 26, 504 (1970).
12. W. D. Henry and B. E. Wilde, An electronic zero resistance ammeter with instantaneous null characteristics, Corrosion, 27, 479 (1971).
13. F. Mansfeld, Results of galvanic corrosion round robin testing, Corrosion, 33, 224 (1977).
14. R. H. Heidersbach, Clarification of the mechanism of the dealloying phenomenon, Corrosion, 24, 38 (1968).
15. R. H. Heidersbach and E. D. Verink, The dezincification of alpha and beta brasses, Corrosion, 28, 397 (1972).
16. R. Natarajan, P. C. Angelo and N. F. George, Dezincification of cartridge brass, Corrosion, 31, 302 (1975).
17. E. E. Langenegger and B. G. Callaghan, Use of an empirical potential shift technique for predicting dezincification rates of α-brasses in chloride media, Corrosion, 28, 245 (1972).
18. L. P. Costas, A potentiostatic method of evaluating the dezincification tendencies of brasses, Corrosion, 30, 167 (1974).
19. V. F. Lucey, Mechanism of dezincification and effect of arsenic, Br. Corr. J., 1, 9 (1965).
20. E. Maahn and R. Blum, Dezincification test for brass based on potentiostatic polarization, Br. Corr. J., 10, 39 (1974).
21. B. I. Dillon, Laboratory dezincification testing, Metals Forum, Australian Inst. Metals, 1, 71 (1978).
22. B. I. Dillon, Dezincification of brasses at Pt. Lincoln, South Australia and correlation with a laboratory dezincification test, Proc. 7th Int. Cong. Metallic Corrosion Rio de Janeiro (1978).

202   D. A. JONES

23. E. D. Verink and P. A. Parrish, Use of Pourbaix diagrams in predicting the susceptibility to dealloying, *Corrosion*, **26**, 214 (1970).
24. R. H. Heidersbach and E. D. Vernik, Evaluation of the tendency for dealloying in metal systems, ASTM STP 516, 303 (1972).
25. R. M. Latanision and R. W. Staehle, Stress corrosion cracking of iron–nickel–chromium alloys, *Proc. Conf. on Fundamentals of Stress Corrosion Cracking*, 214, NACE, Houston (1969).
26. R. W. Staehle, private communication quoted in reference 15.
27. A. Joshi and D. F. Stein, Chemistry of grain boundaries and its relation to intergranular corrosion of austenitic stainless steel, *Corrosion*, **28**, 321 (1972).
28. G. S. Was, R. G. Ballinger, R. M. Latanision and R. M. Pelloux, Second SemiAnnual Progress Report, Research Project 1166–3, Electric Power Research Institute, Palo Alto, CA.
29. R. C. Gibson, H. W. Hayden and J. H. Brophy, Properties of stainless steels with a microduplex structure, *Trans. ASM*, **61**, 85 (1968).
30. F. R. Beckitt, The formation of sigma-phase from delta-ferrite in a stainless steel, *J. Iron and Steel Inst.*, **207**, 632 (1969).
31. T. M. Devine, Mechanism of intergranular corrosion and pitting corrosion of austenitic and duplex 308 stainless steel, *J. Electrochem. Soc.*, **126**, 374 (1979).
32. T. M. Devine, Influence of carbon content and ferrite morphology on the sensitization of duplex stainless steel, *Met. Trans. A*, **11A**, 791, (1980).
33. W. D. France and N. D. Greene, Predicting the intergranular corrosion of austenitic stainless steels, *Corrosion Sci.*, **8**, 9 (1968).
34. M. A. Streicher, The time-factor in potentiostatic studies of intergranular corrosion of austenitic stainless steels, *Corrosion Sci.*, **9**, 53 (1969).
35. W. D. France and N. D. Greene, Some effects of experimental procedures on controlled potential corrosion tests of sensitized austenitic stainless steels, *Corrosion Sci.*, **10**, 379 (1970).
36. W. L. Clarke, V. M. Romero and J. C. Danko, Detection of sensitization in stainless steel using electrochemical techniques, Paper No. 180. CORROSION/77, NACE, (Houston).
37. V. Cihal, A. Desestret, M. Froment and G. H. Wagner, Tests for the study and evaluation of intergranular corrosion in nonoxydizable acids, *Proc. Conf. European Federation on Corrosion*, Paris, 249 (1973).
38. P. Novak, R. Stefec and F. Franz, Testing the susceptibility of stainless steel to intergranular corrosion by a reactivation method, *Corrosion*, **31**, 344 (1975).
39. W. L. Clarke, General Electric Co., Pleasanton, CA94566, private communication.
40. J. S. Armijo, Intergranular corrosion of nonsensitized austenitic stainless steels, *Corrosion*, **21**, 235 (1965).
41. J. S. Armijo, Intergranular corrosion of nonsensitized austenitic stainless steels, *Corrosion*, **24**, 24 (1968).
42. K. T. Aust, J. S. Armijo, E. F. Koch, and J. H. Westbrook, Intergranular corrosion and mechanical properties of austenitic stainless steels, *Trans ASM*, **61**, 270 (1968).

43. D. A. Vermilyea, C. S. Tedmon and D. E. Broecker, Effect of P and Si on the intergranular corrosion of a nickel base alloy, *Corrosion*, **31**, 222 (1975).
44. B. C. Syrett, Erosion corrosion of copper–nickel alloys in sea water and other aqueous environments—a literature review, *Corrosion*, **32**, 243 (1976).
45. J. Postlethwaite and M. W. Hawrylak, Effect of slurry abrasion on the anodic dissolution of iron in pure water, *Corrosion*, **31**, 237 (1975).
46. J. Perkins, K. J. Graham, G. A. Storm, G. Leumer, and R. P. Shark, Flow effects on corrosion of galvanic couples in seawater, *Corrosion*, **35**, 23 (1979).
47. K. D. Efird, Effect of fluid dynamics on the corrosion of Cu-base alloys in sea water, *Corrosion*, **33**, 3 (1977),
48. M. G. Fontana and N. D. Greene, op. cit., p. 72.
49. D. A. Jones, Effect of water chemistry on the erosion-corrosion of aluminum in high temperature high velocity water, *Corrosion*, **37**, 563 (1981).
50. C. Loss and E. Heitz, Mechanisms of erosion-corrosion in fast flowing liquids, *Werkstoffe und Korrosion*, **24**, 38 (1973).
51. G. Bianchi, G. Fiori, P. Longhi and F. Mazza, Horseshoe corrosion of Cu alloys in flowing seawater: mechanism and possibility of cathodic protection of condenser tubes in power stations, *Corrosion*, **34**, 396 (1978).
52. B. F. Brown, Concept of the occluded corrosion cell, *Corrosion*, **26**, 249 (1970).
53. D. F. Taylor, Crevice corrosion of alloy 600 in high temperature aqueous environments, *Corrosion*, **35**, 550 (1979).
54. Z. Szklarska-Smialowska, Pitting corrosion of iron in sodium sulphate, *Corrosion Sci.*, **18**, 97 (1978).
55. R. J. Pickard and N. D. Greene, Electrochemical reactions at paper shielded metal surfaces, *Corrosion*, **30**, 393 (1974).
56. Z. Szklarska-Smialowska, Review of literature on pitting corrosion published since 1960, *Corrosion*, **27**, 223 (1971).
57. W. D. France, Crevice corrosion in metals, *Localized corrosion—cause of metal failure*, ASTM STP 516, 164 (1972).
58. B. E. Wilde and E. Williams, The relevance of accelerated electrochemical pitting tests to long-term pitting and crevice corrosion behavior of stainless steel in marine environments, *J. Electrochem. Soc.*, **118**, 1057 (1971).
59. C. R. Rarey and A. H. Aronson, Pitting corrosion of sensitized ferritic stainless steel, *Corrosion*, **28**, 255 (1972).
60. R. Stefec, F. Franz and A. Holacek, Influence of heat treatment on pitting corrosion of austenitic Cr–Ni–Mo Steels in NaCl solutions, *Werkstoffe und Korrosion*, **30**, 189 (1979).
61. N. Stolica, Pitting corrosion of iron-chromium alloys, *Corrosion Sci*, **9**, 205 (1969).
62. H. P. Leckie and H. H. Uhlig, Environmental factors affecting the critical potential for pitting in 18-8 stainless steel, *J. Electrochem. Soc.*, **113**, 1262 (1966).
63. N. Pessal and C. Liu, Determination of critical pitting potentials of stainless steels in aqueous chloride environments, *Electrochimica Acta*, **16**, 1987 (1971).
64. J. R. Galvele, J. B. Lumsden and R. W. Staehle, Effect of molybdenum on the

pitting potential of high purity 18% Cr ferritic stainless steel, *J. Electro-chem. Soc*, **125**, 1204 (1978).

65. E. A. Lizlovs and A. P. Bond, An evaluation of some electrochemical techniques for the determination of pitting potentials of stainless steel, *Corrosion*, **31**, 219 (1975).

66. B. Mazza, P. Pedeferri, D. Sinigaglia, A. Cigada, G. A. Mondora, G. Re, G. Taccani and D. Wenger, Pitting resistance of a cold-worked commercial austenitic stainless steel in solution simulating seawater, *J. Electrochem. Soc.*, **126**, 2075 (1979).

67. W. D. France and N. D. Greene, Comparison of chemically and elec-trolytically induced pitting corrosion, *Corrosion*, **26**, 1 (1970).

68. B. E. Wilde and E. Williams, On the correspondence between electrochemical and chemical accelerated pitting corrosion tests, *J. Electro-chem. Soc.*, **117**, 775 (1970).

69. B. E. Wilde, Chloride ion adsorption and pit initiation on stainless steels in neutral media, *Passivity and Its Breakdown on Iron Base Alloys*, 129, NACE, Houston (1976).

70. T. Nakama and K. Sasa, Effect of ultrasonic waves on the pitting potentials of 18-8 stainless steel in sodium chloride solution, *Corrosion*, **32**, 283 (1976).

71. F. Franz and P. Novak, Effect of rotation on the pitting corrosion of aluminum electrodes, *Localized Corrosion*, NACE-3, NACE, Houston, 576 (1974).

72. F. Mansfeld and J. V. Kenkel, Effect of rotation on pitting behavior of aluminum and stainless steel, *Corrosion*, **35**, 43 (1979).

73. M. Janik-Czakor, A. Szummer and Z. Szklarska-Smialowska, Electron microprobe investigation of processes leading to the nucleation of pits on iron, *Corrosion Sci.*, **15**, 775 (1975).

74. J. Augustynski, F. Dalard and J. C. Sohm, Anodic oxidation of zinc in weakly basic media, *Corrosion Sci.*, **12**, 713 (1972).

75. H. H. Strehblow and M. B. Ives, On the electrochemical conditions within small pits, *Corrosion Sci.*, **16**, 317 (1976).

76. Z. Szklarska-Smialowska, H. Viefhaus and M. Janik-Czakor, Electron microscopy analysis of in-depth profiles of passive films formed on iron in chloride containing solutions, *Corrosion Sci.*, **16**, 649 (1976).

77. J. B. Lumsden and R. W. Staehle, Application of Auger electron spec-troscopy to the determination of the composition of passive films on Type 316 SS, *Scripta Met.*, **6**, 1205 (1972).

78. K. E. Heusler and L. Fischer, Kinetics of pit initiation at passive iron, *Werkstoffe und Korrosion*, **27**, 551, 697, 788 (1976).

79. Z. Szklarska-Smialowska, Influence of sulfide inclusions on the pitting corrosion of steels, *Corrosion*, **28**, 388 (1972).

80. M. Janik-Czakor, A. Szummer, and Z. Szklarska-Smialowska, Effect of sulfur and manganese on nucleation of corrosion pits in iron, *Br. Corr. J.*, **7**, 90 (1972).

81. B. Forchhammer and H. J. Engell, Pitting corrosion of passive austenitic Cr–Ni steels in neutral chloride solutions, *Werkstoffe und Korrosion*, **20**, 1 (1969).

82. B. E. Wilde and J. S. Armijo, Influence of sulfur on the corrosion resistance of austenitic stainless steel, *Corrosion*, **23**, 208 (1967).

83. Z. Szklarska-Smialowska, A. Szummer, and M. Janik-Czakor, Electron microprobe study of the effect of sulfide inclusions on the nucleation of corrosion pits in stainless steels, *Br. Corr. J.*, **5**, 159 (1970).
84. P. E. Manning, D. J. Duquette and W. F. Savage, Effect of test method and surface condition on pitting potential of single and duplex phase 304L stainless steel, *Corrosion*, **35**, 151 (1979).
85. R. Stefec and F. Franz, A study of the pitting corrosion of cold worked stainless steel, *Corrosion Sci.*, **18**, 161 (1978).
86. Z. Szklarska-Smialowska and M. Janik-Czakor, Electrochemical investigation of the nucleation and propagation of pits in iron chromium alloys, *Br. Corr. J.*, **4**, 138 (1969).
87. A. Randak and F. W. Trautes, Effect of austenite stability of 18/8 chromium nickel steels on their cold working and corrosion properties, *Werkstoffe und Korrosion*, **21**, 97 (1970).
88. R. B. Diegle, Localized corrosion of amorphous Fe-Ni-Cr-P-B alloys, *Corrosion*, **35**, 250 (1979).
89. T. M. Devine, Anodic polarization and localized corrosion behavior of amorphous $Ni_{35}Fe_{30}Cr_{15}P_{14}B_6$ in near neutral and acidic chloride solutions, *J. Electrochem. Soc.*, **124**, 38 (1977).
90. A. Baghdasarian and S. F. Ravitz, Corrosion resistance of TRIP steels, *Corrosion*, **31**, 182 (1975).
91. S. G. Ecklund, On the initiation of crevice corrosion on stainless steel, *J. Electrochem. Soc.*, **123**, 170 (1976).
92. L. J. Gainer and G. R. Wallwork, Effect of alloy element additions on the pitting of mild steel, *Corrosion*, submitted for publication.
93. L. J. Gainer and G. R. Wallwork, The effect of non-metallic inclusions on the pitting of mild steel, *Corrosion*, **35**, 435 (1979).
94. B. C. Syrett, D. D. Macdonald and H. Shih, Pitting resistance of engineering materials in geothermal brines—I. Low salinity brine, *Corrosion*, **36**, 130 (1980).
95. B. E. Wilde and N. D. Greene, The variable corrosion resistance of 18Cr–8Ni stainless steels: behavior of commercial alloys, *Corrosion*, **25**, 300 (1969).
96. J. R. Galvele, Transport processes and the mechanism of pitting of metals, *J. Electrochem. Soc.*, **123**, 464 (1976).
97. J. R. Galvele, Present state of understanding of the breakdown of passivity and repassivation, *Fourth Int. Symposium on Passivity*, Arlie, Virginia (1977).
98. M. Pourbaix, Significance of protection potential in pitting and intergranular corrosion, *Corrosion*, **26**, 431 (1970).
99. M. Pourbaix, Characteristics of localized corrosion of steel in chloride solutions, *Corrosion*, **27**, 449 (1971).
100. T. Suzuki and Y. Kitamura, Critical potential for growth of localized corrosion of stainless steel in chloride media, *Corrosion*, **28**, 1 (1972).
101. B. E. Wilde and E. Williams, The use of current/voltage curves for the study of localized corrosion and passivity breakdown on stainless steels in chloride media, *Electrochimica Acta*, **16**, 1971.
102. B. E. Wilde, A critical appraisal of some popular laboratory electrochemical tests for predicting the localized corrosion resistance of stainless alloys in sea water, *Corrosion*, **28**, 283 (1972).

103. E. D. Verink and M. Pourbaix, Use of electrochemical hysteresis techniques in developing alloys for saline exposures, *Corrosion*, **27**, 495 (1971).
104. D. A. Jones and N. D. Greene, Electrochemical detection of localized corrosion, *Corrosion*, **25**, 367 (1969).
105. E. J. Sutow and D. W. Jones, A crevice corrosion cell configuration, *J. Dental Res.*, **58**, 1358 (1974).
106. J. M. Drugli and E. Bardal, A short duration test method for prediction of crevice corrosion rates on stainless steels, *Corrosion*, **34**, 419 (1978).
107. R. M. Kain, Localized corrosion behavior in natural seawater: a comparison of electrochemical and crevice testing of stainless alloys, Paper No. 74, Corrosion /80, NACE, Houston (1980).
108. B. C. Syrett, PPR curves—a new method of assessing pitting corrosion resistance, *Corrosion*, **33**, 221 (1977).
109. B. C. Syrett and S. S. Wing, Pitting resistance of new and conventional orthopedic implant materials—effect of metallurgical condition, *Corrosion*, **34**, 138 (1978).
110. J. F. Bates, Cathodic protection to prevent crevice corrosion of stainless steels in halide media, *Corrosion*, **29**, 28 (1973).
111. D. A. Jones and B. E. Wilde, Galvanic reactions during localized corrosion on stainless steel, *Corrosion Sci.*, **18**, 631 (1978).
112. I. L. Rozenfeld and I. S. Danilov, Electrochemical aspects of pitting corrosion, *Corrosion Sci.*, **7**, 129 (1967).
113. T. Suzuki, M. Yamabe and Y. Kitamura, Composition of the anolyte within the pit anode of austenitic stainless steels in chloride solution, *Corrosion*, **29**, 18 (1973).
114. J. Mankowski and Z. Szklarska-Smialowska, Studies on accumulation of chloride ions in pits growing during anodic polarization, *Corrosion Sci.*, **15**, 493 (1975).
115. Y. Kitamura and T. Suzuki, Pit growth in stainless steel and its prevention by an artificial pit method, *Proc. 4th Int. Cong. Metallic Corrosion*, 716, NACE, Houston (1972).
116. H. W. Pickering and R. P. Frankenthal, On the mechanism of localized corrosion of iron and steel-I. Electrochemical studies, *J. Electrochem. Soc.*, **119**, 1297 (1972).
117. H. H. Strehblow and M. B. Ives, On the electrochemical conditions within small pits, *Corrosion Sci.*, **16**, 317 (1976).
118. K. Nisancioglu and H. Holtan, Correlation of the protection potential and the ohmic potential drop, *Electrochimica Acta*, **23**, 251 (1978).
119. E. D. Verink, T. S. Lee and R. L. Cusumano, Influence of prior electrochemical history on the propagation of localized corrosion, *Corrosion*, **28**, 348 (1972).
120. N. Pessall and J. I. Nurminen, Development of ferritic stainless steels for use in desalination plants, *Corrosion*, **30**, 381 (1974).
121. E. D. Verink, K. K. Starr, and J. M. Bowers, Chemistry of crevice corrosion as observed in certain copper–nickel and iron–chromium alloys, *Corrosion*, **32**, 60 (1976).
122. K. Nisancioglu and H. Holtan, The protection potential of aluminum, *Corrosion Sci.*, **18**, 1011 (1978).

123. S. B. de Wexler and J. R. Galvele, Anodic behavior of aluminum straining and a mechanism for pitting, *J. Electrochem. Soc.*, **121**, 1271 (1974).
124. D. A. Jones, The application of electrode kinetics to the theory and practice of cathodic protection, *Corrosion Sci.*, **11**, 439 (1971).
125. S. M. Laskowski and M. J. Wooten, Use of neutralizers to control denting in steam generators. Presented at the American Power Conference, April 1979, Chicago.
126. R. Garnsey, Corrosion of PWR steam generators, *Nuclear Energy*, **18**, 117 (1979).
127. E. C. Potter and G. M. W. Mann, Fast linear growth of magnetite on mild steel in high temperature aqueous conditions, *Br. Corr. J.*, **1**, 26 (1965).
128. H. H. Uhlig, op. cit., p. 130.

Chapter 5

CORROSION FATIGUE

J. Congleton and I. H. Craig
*Department of Metallurgy and Engineering Materials,
The University of Newcastle upon Tyne, UK*

1. INTRODUCTION

This review of corrosion fatigue in metals and alloys adopts a starting point at the review by Gilbert[1] in which the pre-1956 knowledge was comprehensively summarized. There are difficulties in deciding what material should be included because of the nebulous boundaries of the subject. In general, air fatigue data are used as a basis for comparisons although it is well known that laboratory air is certainly not 'inert' with respect to fatigue crack growth. For instance, an iron-base super alloy shows significant reductions in low cycle fatigue strength in air at 593 °C, compared to data *in vacuo*, that are frequency dependent.[2] Similarly, reductions in fatigue strengths in air compared with the strengths *in vacuo* have been observed, for example in copper,[3] 70:30 brass,[3] molybdenum,[4] lead,[5] aluminium,[6] and aluminium alloys.[7,8] With the advent of fracture mechanics more meaningful comparisons could be made between crack propagation rates in air and *in vacuo*. Most of this latter work has been carried out on aluminium and its alloys reflecting the importance placed upon a knowledge of the crack propagation rates of these materials for aerospace applications. In the majority of these investigations[6,7,9-15] fatigue crack propagation rates *in vacuo* have been observed to be lower than in air at lower stress levels. However, the effects of gaseous environments are worthy of a separate review and most of the further comment will be concerned with the effect of aqueous environments.

Gilbert's review was quite extensive and highlighted several general aspects of corrosion fatigue that are still valid and worth repeating at the outset. For instance, attention was drawn to the problems of data collection especially concerning the possible permutations of important test variables. The role of time *per se* and its interrelation with the direct corrosive action of the environment and the testing frequency was firmly emphasized. Thus, at low cyclic frequencies, the total test time may be the dominant factor rather than the number of stress cycles. Gilbert also indicated the importance of stress amplitude and mean stress level, although the now commonly employed R value, defined as the ratio minimum stress/maximum stress, was not used. Testing temperature, specimen and loading types and specimen surface condition were all commented upon and the review contained a factual account of a considerable amount of corrosion fatigue data in terms of conventional $S-N$ information.

The major effect of corrosion on the form of the $S-N$ curve is that fatigue limits in the true sense are removed so that a finite lifetime at any stress tends to exist providing cyclic loading is continued for long enough in the environment. It is worth noting that fatigue life is often prolonged at high stress amplitudes by the presence of certain environments, so that corrosion may lead to an $S-N$ curve of the form of either A or B in Fig. 1. For some materials the environment is aggressive enough at high stress levels to retard initiation hence prolonging the fatigue life whereas in other systems the whole $S-N$ curve is moved to lower values in the aggressive environment. Similar effects occur for materials that do not show true fatigue limits in their air test data.

Gilbert also made the following points:

1. Corrosion fatigue cracks are usually transcrystalline though in a minority of cases they may be intercrystalline.

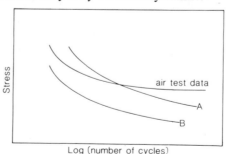

FIG. 1. $S-N$ curves for air and corrosion fatigue tests, schematic. A—corrosion fatigue showing retarded initiation at high stress, B—corrosion fatigue giving a general lowering of fatigue strength.

2. Corrosion fatigue often produces branched cracks rather than the single fronted cracks characteristic of air fatigue.
3. Corrosion fatigue in immersed conditions is an electrochemical phenomenon because it can sometimes be prevented by cathodic protection or by adding inhibitors to the corroding solution.
4. The corrosion fatigue limit, assessed usually as a stress that gives failure in a specific lifetime, is insensitive to metallurgical conditions and shows little correlation with strength, in contrast to the generally good correlation between the tensile strength and air fatigue strength of many alloys. This point is clearly illustrated in Fig. 2, due to Kitegawa[16] from data of McAdam.[17]

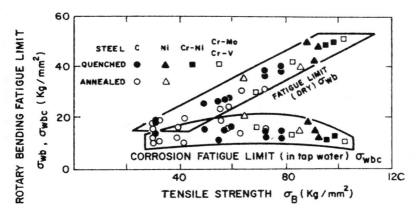

FIG. 2. Fatigue data for carbon and alloy steels. From Kitegawa, Ref. 16.

The early work of Gough and Sopwith[18,19] had clearly demonstrated that air does decrease the fatigue strength relative to tests *in vacuo* and that for copper the combined presence of water vapour and oxygen was the cause of the atmospheric effects. For tests in aqueous solutions also, the presence of oxygen had been demonstrated as necessary to induce corrosion fatigue for certain systems.[20] However, oxygen as such is not an absolute requirement for the inducement of corrosion fatigue and the possible role of hydrogen was referred to by Gilbert.[1]

It was also clear by 1956 that the initiation and propagation stages of corrosion fatigue were somewhat different. Gilbert reviewed the pre-pitting tests of McAdam and discussed the possible role of pitting in crack initiation. Pits were known to form in some systems, eventually leading to fatigue cracks, and isolated pits were recognized as more effective than associated or multiple pits. This was considered to result from differences in stress concentration factors for various pit geometries. Clearly, Gilbert's review provided the foundations on which development of more quantitative research into corrosion fatigue could be built both with respect to electrochemical studies and to fatigue strength data collection. This has occurred and has been aided by the advent of fracture mechanics.

In recent years there have been several conferences with sections, and in some cases the complete proceedings, devoted to the subject of corrosion fatigue. Some of these publications are given as Refs 21 to 25. Additionally, several review articles have been published either as separate papers or chapters in specialist books.[26-30] Thus, it seems illogical to provide here a mere regurgitation of what is already quite extensively in print. Rather, this review is aimed at highlighting some aspects of corrosion fatigue that still require detailed research. Extensive experimental programmes are still required because the nature of the subject is such that a universal explanation of corrosion fatigue is unlikely ever to be forthcoming. Many mechanisms are possible for a corrosive environment to enhance or even retard fatigue. The corrosive nature of the liquid may indeed be very mild but significant changes in fatigue cracking may occur.

It is well known that the initiation of fatigue is controlled by a different mechanism to fatigue crack growth, in the sense that initiation can be induced by localized plastic deformation unaided by any geometric stress concentrator and it is the relative values of applied stress and yield strength that are important. For fatigue crack propagation, fracture toughness parameters become dominant in determining the rate of crack extension, and the crack tip stress intensity factor becomes the controlling parameter.

2. THE INITIATION OF CORROSION FATIGUE CRACKS

The work of McAdam[17] on pre-corroded specimens clearly indicated that pitting was likely to be associated with the early initiation of fatigue cracks in a corrosion fatigue situation. Typically, 90% of the air fatigue life of smooth polished specimens may be associated with the initiation of fatigue cracks and only 10% with their growth. However, machined notches and/or the presence of an appropriate environment can drastically reduce fatigue life by reducing the time for initiation to about only

10% of the total lifetime. Notches act by concentrating plastic strain in the high stress region near their tips, thus making crack nucleation easier. It is to be expected therefore, that if a corrosive environment creates pits, the latter will act as stress raisers and induce earlier crack initiation. McAdam indeed showed this and demonstrated that whereas prepitting by corrosion reduced the air fatigue life, prepitting then testing in the same corrosive environment caused a further reduction of fatigue strength. He also showed that the major reduction of fatigue strength occurred after the first few days of pre-corrosion; extended corrosion times up to 300 days caused little further reduction of fatigue strength. This is understandable because although a small pit produces a significant stress intensification, by a factor of about 2.2 for an hemispherical pit[31,32] the stress intensification factor will remain the same if the pit remains hemispherical as it deepens, and increases to only about 3.5 if it deepens and becomes hyperbolic in shape. Thus, high stress concentration factors will only arise for narrow, sharp tipped, relatively long pits which are probably geometrically indistinguishable from cracks.

However, corrosion pits are not essential for easier crack nucleation as was shown by Simnad and Evans[33] and Duquette and Uhlig,[34] who tested steel in acid solutions and obtained corrosion fatigue in the absence of pitting corrosion conditions. Laird and Duquette[35] have discussed the necessity or otherwise of pitting for the initiation of corrosion fatigue and concluded that for steels, perhaps pits observed at failure were not the cause of corrosion fatigue but rather the result of it. However, this conclusion though justified in some cases is unjustified in others. The truth is more likely to be that several mechanisms of initiation are possible and the one with the fastest kinetics will dominate. For instance, in the air fatigue of polycrystalline aluminium, fatigue crack initiation occurs at persistent slip bands at low stress but by grain boundary initiation at high stresses. For corrosion fatigue of polycrystalline aluminium in 3.5% NaCl at pH 7, crack initiation occurs at persistent slip bands rather than at grain boundaries and over a wider range of stresses than is the case for experiments in air. The reason for this is that the chloride environment produces pitting at the persistent slip bands and so encourages these as sites for crack nucleation rather than the grain boundaries.

The pH of 3.5% NaCl has a significant effect on the initiation of corrosion fatigue in aluminium[36] as shown in Fig. 3. Especially at low stress amplitudes, initiation is much more rapid in pH 7 and pH 8 solutions than at pH 10. Indeed, initiation tends to be retarded relative

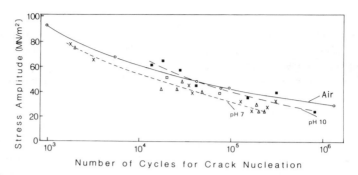

FIG. 3. Corrosion fatigue crack initiation data for aluminium in 3.5% sodium chloride.

to air tests at pH 10 at high stress amplitudes. The reason for this is apparent from examination of sectioned specimens. At pH 7, the initiated cracks are sharp (Fig. 4(a)), and this acute geometry persists as the cracks propagate. At pH 10, however, general surface dissolution occurs and crystallographically related lateral spreading of pits at persistent slip bands occurs, giving very blunt-shaped crevices which are unsuitable sites for crack initiation (Fig. 4(b)), so the initiation of fatigue cracks in pH 10 solutions tends to be at grain boundaries where strain incompatibility leads to relatively sharp profiled cracks.

The reason for the effect of pH is the variation in solubility of Al_2O_3, as shown in Fig. 5.[37] As Al_2O_3 is relatively insoluble at pH 7, conditions for localized attack at disruptions in the surface oxide, coupled with favourable conditions for passivation of the crack sides, cause the development of sharp cracks. The three orders of magnitude increase in solubility of Al_2O_3 in moving from pH 7 to pH 10 obviously removes the tendency for passivation and lateral dissolution occurs at preferential corrosion sites, i.e. persistent slip bands. Indeed, the general corrosion observed at pH 10 retards initiation relative to air tests, especially at high stress amplitudes. This is a common feature of a considerable amount of corrosion fatigue data and is indicated schematically by line A in Fig. 1.

Initiation data for a BS 4360, 50D steel[38] are presented in Fig. 6, which, in contrast to the situation in aluminium just mentioned, shows that crack initiation is more rapid at pH 10 than at pH 7 and that the S–N curve as a whole is moved to lower stress limits. For steel, increasing the pH from 7 to 10 decreases the general corrosion, i.e. the opposite effect to aluminium in 3.5% NaCl. In the 50D steel crack nucleation was

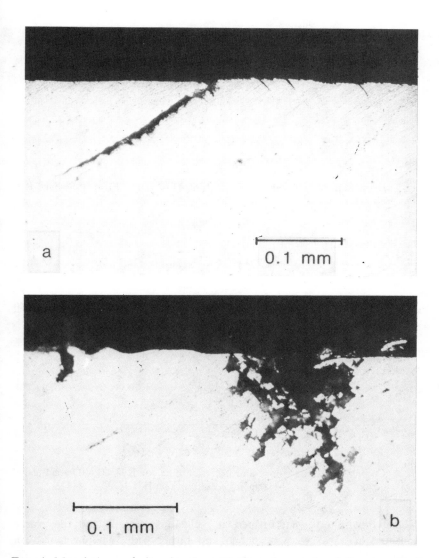

FIG. 4. Morphology of sites for the initiation of corrosion fatigue cracks in aluminium, (a) pH 7, 3.5% sodium chloride, (b) pH 10, 3.5% sodium chloride.

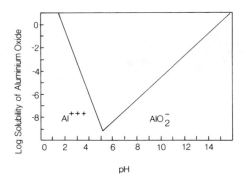

FIG. 5. Dependence of solubility of Al_2O_3 on pH of water. From Pourbaix, Ref. 37.

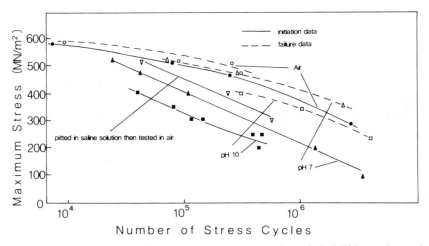

FIG. 6. Corrosion fatigue failure and initiation data for BS4360/50D steel tested at 20 Hz in 3.5% sodium chloride.

associated with pits that formed at inclusions (Fig. 7) and the cracks generated at pH 7 were branched and rather wide, whereas narrower unbranched cracks nucleated at pH 10. The situation is comparable to that existing for aluminium in that electrochemical conditions leading to localized attack and the generation of relatively sharp cracks give the maximum reduction in time for initiation whereas general corrosion, which encourages blunting of cracks, delays initiation and may even

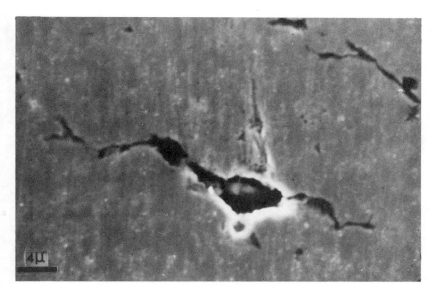

FIG. 7. Pit forming at inclusion in BS4360/50D steel tested in pH 10, 3.5% sodium chloride.

enhance fatigue life, though this latter effect is usually for high alternating stresses where lifetime is in any case relatively short. However, this does indicate a dependence on stress level as well as electrochemical factors. For instance, Fig. 8 shows data for the rates of pit deepening under static load and initial crack growth rates for cyclically loaded 50D steel when immersed in pH 10, 3.5% NaCl. Clearly, the rate of pit growth is enhanced at static stresses above the yield strength. For cyclic loading, cracks readily form at pits so it is difficult to obtain data on pit growth under corrosion fatigue conditions for direct comparison, so it is therefore impossible to decide to what extent cyclic loading enhances pit growth rates.

Thus, there is factual evidence that, both for aluminium and for steel in 3.5% NaCl solutions, corrosion fatigue crack initiation is dependent upon the details of pit formation. However, other possibilities exist. For instance, grain boundary attack, hydrogen assisted cracking and localized attack at emerging slip steps that break a passivating layer are all possible mechanisms for enhanced fatigue crack initiation, with preferential attack at emergent slip lines having close similarities to the attack of the persistent slip lines in aluminium already discussed.

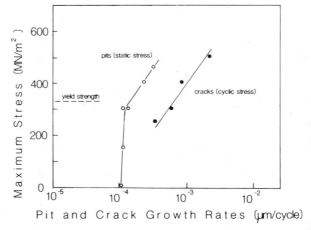

FIG. 8. Pit growth and initial crack growth rates for BS4360/50D steel tested in pH 10, 3.5% sodium chloride.

The above examples support the final remark made by Laird and Duquette in their review paper[35] that 'the exact machanism (or mechanisms) of fatigue crack initiation in the presence of both inert and corrosive environments remains elusive and rather elaborate experimentation is required before satisfactory results can be developed'. Certainly, research into the initiation of corrosion fatigue is difficult for two major reasons. First, the definition of initiation is arbitrary, i.e. when can an intrusion or deep pit be described as a crack? Second, corrosion fatigue is characterized by the development of many cracks rather than a single crack (Fig. 9) so that obtaining quantitative data on crack initiation is a tedious and lengthy task. This multiplicity of cracking that increases as load cycling continues means that the stress concentration at a single crack tip is probably negligible unless it is relatively long compared to its immediate neighbours. Thus, microstructural factors may significantly influence corrosion fatigue life according to whether they impede or enhance the probability of a single crack rapidly out-running its neighbours. Also, because initiation is so abundant and relatively easy for corrosion fatigue situation, designers frequently apply finite lifetime concepts to components rather than consider the possible existence of a reduced fatigue limit. Johnson et al.[39] have considered specific applications of this philosophy for offshore structures and nuclear pressure vessels.

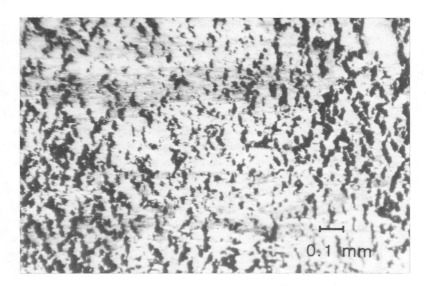

FIG. 9. Multiple corrosion fatigue crack initiation for BS4360/50D steel in pH 10, 3.5% sodium chloride.

3. THE APPLICATION OF FRACTURE MECHANICS TO CORROSION FATIGUE STUDIES

The detailed applications of fracture mechanics to cracked components are extensively covered in the literature and such books as 'Fracture of Structural Materials',[40] Fundamentals of Fracture Mechanics[41] and Elementary Engineering Fracture Mechanics,[42] among others, cover the subject well. Thus, a brief statement of the relevant essentials is sufficient here.

The tensile stress ahead of a sharp crack is given[43] by

$$\sigma_{yy} = \frac{K}{(2\pi r)^{\frac{1}{2}}}$$

where σ_{yy} is the stress normal to the plane of the crack and r the distance measured from the crack tip. For a crack in an infinite or semi-infinite plate the stress intensity factor, K, has the limiting value $\pi^{\frac{1}{2}}\sigma_{app}a^{\frac{1}{2}}$ at $r=0$ providing the crack tip plastic zone is very small. Here, a is the semi-length of an internal crack or the full length of an edge crack and σ_{app} is the applied nominal tensile stress. Thus, the stress field just ahead of a sharp crack can be characterized by K and appropriate corrections can be made to allow for finite-sized specimens and various crack geometries. If the crack faces are displaced normal to the crack plane during fracture the relevant K is designated K_I (mode I cracking) and an additional subscript c is used to denote a critical K level (say for unstable crack propagation). Hence, providing certain test geometry restrictions are conformed to, a material parameter, K_{Ic}, can be evaluated experimentally that describes the toughness of precracked material.

There has been a significant move in recent years towards the collection of fatigue data using the fracture mechanics approach because in principle the data can be made specimen independent in the sense that tests on different geometry specimens of the same material should produce the same crack growth versus K data. The actual parameters of importance in corrosion fatigue data collection will be K_{min}, K_{max}, the minimum and maximum values of K in a load cycle, and $\Delta K = K_{max} - K_{min}$. These are functions of crack length as well as load so they may or may not vary throughout a test depending upon the details of the specimen type and the loading schedule. Additionally, as previously indicated, the mean stress may also be important. This is normally expressed in terms of the R ratio, where $R = \sigma_{min}/\sigma_{max}$. Hence, R is positive for tension–tension tests, zero for reversed bending with a minimum stress of zero and negative for compression–tension tests.

As $K = \pi^{\frac{1}{2}} \sigma_{app} a^{\frac{1}{2}} f$ (specimen size, crack size, yield strength) it is clear that

$$1 - R = 1 - \frac{\sigma_{min}}{\sigma_{max}} = \frac{\Delta K}{K_{max}}$$

Also

$$1 + R = \frac{2\bar{K}}{K_{max}}$$

where $\bar{K}$ is the mean K value.

Thus, R, ΔK and K_{max} are not independent test variables and care is necessary when interpreting data where all of these parameters are apparently varied, or in comparing data plotted with different coordinate axes.

Another point of some significance for corrosion fatigue testing is the phenomenon of crack closure as described by Elber.[44] Briefly, in an elastic–plastic medium that undergoes limited plastic deformation at the crack tip during loading, it is possible for the crack faces to impinge during unloading before the tensile load is completely removed. For thin plates, Elber suggested that this reduced the effective ΔK during fatigue and has proposed the empirical relationship $\Delta K_{\text{effective}} = (0.5 + 0.4R)\Delta K_{\text{apparent}}$ based on results for aluminium. It is perhaps fair to say that this result is pertinent to some situations, but crack closure effects are yet far from being fully understood. This is not surprising because the details of what conditions pertain to a deforming crack tip in a corrosive environment are extremely difficult to investigate as will be emphasized later.

4. CORROSION FATIGUE CRACK GROWTH

Because fracture mechanics allows conditions for equivalent stressed states at a crack tip to be evaluated for many different specimen geometries and loading conditions, it is convenient to relate crack growth rate data to the crack tip stress intensity factor amplitude during fatigue. Crack growth data are also of practical significance for corrosion fatigue because many structures must be assumed to have cracks in existence when they are commissioned or will develop these relatively quickly and their service life will be determined by how fast these cracks grow under the relatively low frequency loading that is most pertinent to corrosion fatigue.

In any crack growth situation, the most dominant cracking mechanism is likely to control the growth rate. Thus, for high frequency loading, air fatigue growth rates tend to be unaltered by the presence of an aqueous corrosive environment simply because there is insufficient time per cycle for corrosive effects to influence significantly fatigue crack growth. This is clearly shown in Fig. 10 which presents data due to Barsom[45] indicating the marked reduction of enhanced crack growth due to corrosion as the test frequency is increased to 600 cpm (10 Hz).

The slow crack growth that may occur in the presence of nominally static stresses plus certain corrosive environments results from mechanisms similar to those involved in explaining corrosion fatigue and the influence upon the latter of stress corrosion cracking has received some attention, frequently along the lines proposed by Wei and Simmons.[46] Briefly, for many systems, air fatigue crack growth occurs only above a threshold stress intensity factor amplitude, ΔK_{th}. The growth per cycle first increases rapidly with increasing ΔK above ΔK_{th}

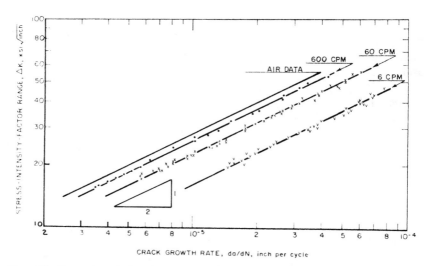

FIG. 10. Corrosion fatigue crack growth data as a function of test frequency.
From Barsom, Ref. 45.

then enters a region where log da/dN is linearly dependent upon log ΔK and finally becomes very rapid as K_{max} approaches K_c or K_{Ic}. This is shown schematically in Fig. 11(a). In contrast, for many systems that stress-corrode, there is a definable stress intensity factor below which crack growth will not occur, termed K_{ISCC}, and the log (crack growth rate) versus log K curve first rises steeply and frequently reaches a plateau value that is independent of K, then rises steeply again when K approaches K_c or K_{Ic} as indicated in Fig. 11(b).

There are thus three possible forms for log da/dN versus log ΔK environment-assisted fatigue curves, as shown in Fig. 11(c), (d) and (e), that depend upon whether or not there is a synergistic effect of the corrosive environment and the alternating stress that is independent of any stress corrosion mechanism and on what value the minimum stress intensity for stress corrosion, K_{ISCC}, if it exists, has in relation to ΔK_{th}. Now the contribution to crack growth from stress corrosion depends upon K_{max} whereas ΔK is the controlling parameter for fatigue crack growth and as $\Delta K/K_{max} = 1 - R$ it is not surprising that many corrosion fatigue data show an R dependence whereas air fatigue data are generally independent of R value. Many data have been collected on the basis of log da/dN versus log ΔK plots at various frequencies and at various R values. It is difficult to collate this information in a compact form

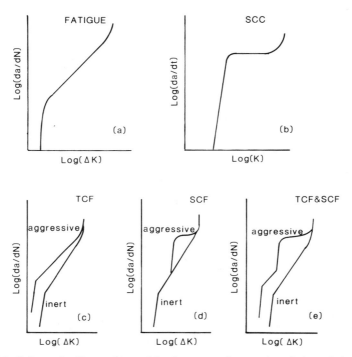

FIG. 11. Schematic illustrations of basic types of corrosion fatigue behaviour. After Refs 46 and 75: SCC = stress corrosion cracking; TCF = true corrosion fatigue; SCF = stress corrosion fatigue.

because of the range of complexity of the log da/dN versus log ΔK plots, but some generalization is possible as indicated below.

In 1963, Paris and Erdogan[47] suggested that at least part of fatigue crack growth conformed to a relationship

$$\frac{da}{dN} = C\Delta K^n \qquad (1)$$

The exponent n varies considerably and there is a systematic variation of C with n. Tanaka and Matsuoka[48] and later Tanaka[49] showed that $C = A/\Delta K_0$ for many steels, where A and ΔK_0 were considered material constants. A similar expression to eqn (1) can be applied to corrosion fatigue data for a range of materials in various environments[50] as, for instance, with the data in Tables 1 to 4 which plot as shown in Fig. 12. Each point in Fig. 12 represents the Paris law portion of some fatigue

TABLE 1
PARIS CONSTANT AND PARIS EXPONENT VALUES FOR LIGHT ALLOYS

Alloy	Environment	log_{10} C	n	Ref.
	Air	−10.75	3.53	
	Seawater, free corrosion	−10.47	3.41	
5086–H116	Seawater −0.75 V	−10.09	3.16	110
	Seawater −1.3 V	−10.82	3.69	
	Seawater −1.4 V	−11.13	3.76	
	Air	−10.78	3.53	
	Seawater, free corrosion	−9.09	2.26	
5086–H117	Seawater −0.75 V	−9.55	3.05	110
	Seawater −1.3 V	−10.50	3.30	
	Seawater −1.3 V	−11.43	3.95	
	Seawater −1.4 V	−11.47	3.95	
	Air	−10.67	3.53	
	Air	−10.90	3.68	
	Seawater, free corrosion	−11.16	4.20	
5456–H116	Seawater −0.75 V	−10.95	3.79	110
	Seawater −1.3 V	−10.77	4.25	
	Seawater −1.3 V	−10.95	3.70	
	Seawater −1.4 V	−13.06	5.10	
	Air	−11.01	3.69	
	Seawater, free corrosion	−9.85	2.97	
5456–H117	Seawater − 0.75 V	−9.43	2.74	
	Seawater − 1.3 V	−11.81	4.11	
	Seawater − 1.4 V	−15.02	6.31	
	Seawater − 1.5 V	−15.66	6.86	
	Air	−9.91	3.0 ($\Delta K < 28$ MN m$^{-3/2}$)	
	Seawater − 0.75 V	−12.91	5.0 ($\Delta K > 28$ MN m$^{-3/2}$)	
		−9.91	3.0 ($\Delta K < 28$ MN m$^{-3/2}$)	
5456–H116		−12.91	5.0 ($\Delta K > 28$ MN m$^{-3/2}$)	111
	Seawater − 0.95	−11.3	4.4	
	Seawater − 1.3	−9.91	3.0	
		−12.91	5.0	
	Argon	−10.07	2.73	
	H$_2$O 1 Pa pressure	−10.00	2.73	
	2 Pa pressure	−10.00	2.73	
2219–T851	3.3 Pa pressure	−9.95	2.73	112
	4.7 Pa pressure	−9.70	2.73	
	6.9 Pa pressure	−9.65	2.73	
	26.6 Pa pressure	−9.60	2.73	
Mg–Al	Dry argon	−10.4	2.37	
ZK60A–T5	5 M, Nacl, 23 °C	−10.19	2.37	113
	5 M, NaBr, 23 °C	−9.85	2.37	
Ti–6Al–2Nb–	Air			111
1Ta–0.8Mo	Seawater	−10.81	3.0	
	Dry argon	−12.2	4.3	
Ti–6Al–4V	Distilled water	−11.52	4.3	113
	0.6 M KCl	−10.32	3.3	

TABLE 2
PARIS CONSTANT AND PARIS EXPONENT VALUES
FOR CONSTRUCTION STEELS

Alloy	Environment	$log_{10}\ C$	n	Ref.
A558	3% NaCl	−11.69	3.5	29
A514 grade F	3% NaCl	−10.74	2.7	29
A36	3% NaCl	−12.37	4.0	29
A588 grade A	3% NaCl	−10.90	3.15	29
A36	Air; 3% NaCl, H_2O	−11.3	3.25	29
A514 grade F	3% NaCl	−10.31	2.5	29
A533B	Distilled water	−12.79	4.0	29
A36	3% NaCl	−10.92	3.0	29
A514 grade E	3% NaCl	−10.27	2.6	29
A514 grade F	3% NaCl	−10.24	2.5	29
50D	Air $R=0.1$	ub† −11.37		
		mean −11.54	3.23	114
50D	Seawater, $R=0.7$	ub− 9.49	2.53	114
50D	Seawater, $R=0.1$	ub −11.37	3.23	114
	Water, 288 °C	−11.69	3.3	
	93 °C	−11.12	3.1	
	93 °C	−10.75	3.2	
A508-2	288 °C	−10.97	3.6	115
	93 °C	−9.87	2.7	
	288 °C	−8.03	1.4	
	PWR	−10.96	3.73	
	PWR	ub −13.09		
A533B		lb† −13.53	4.24	116
	H_2S, 93 °C	−20.45	10.33	
	H_2S, RT	−11.47	4.11	
A533B	Water 288 °C	−9.8	2.15	117
	Water, 288 °C—irradiated	−19.5	9.6	

†ub—upper bound value
 lb—lower bound value

TABLE 3
PARIS CONSTANT AND PARIS EXPQNENT VALUES FOR ALLOY STEELS

Alloy	Environment		$log_{10}\ C$	n	Ref.
	Air		−11.07	2.78	
	Distilled water		−10.47	2.45	
	3.5% NaCl,	5°C, 2.5 Hz	−10.17	2.69	
		5°C, 0.5 Hz	−9.59	2.5	
		5°C, 0.05 Hz	−10.51	2.65	
		25°C, 2.5 Hz	−9.97	2.55	
		0.5 Hz	−9.40	1.9	
HY130		0.05 Hz	−9.48	2.04	118
		0.005 Hz	−9.85	2.41	
		55 °C, 2.5 Hz	−9.87	2.19	

TABLE 3 (continued)

Alloy	Environment		log_{10} C	n	Ref.
		0.5 Hz	−9.33	1.98	
		0.05 Hz	−8.49	1.62	
	85°C,	2.5 Hz	−9.17	1.84	
		0.5 Hz	−8.76	1.66	
		0.05 Hz	−8.23	1.53	
	Air		−10.74	2.7	
			−10.02	2.24	
			−10.25	2.4	
HY130			−10.23	2.36	119
			−9.93	2.19	
			−9.48	1.91	
			−10.21	2.26	
	Air		−9.02	2.15	
HY130	Seawater −0.665 AgCl		−9.20	1.86	111
	−1.050 AgCl		−8.60	1.64	
	−1.050 AgCl		−8.55	1.71	
HY130, 10Ni–Cr–	Air		−9.58	2	29
Mo–V	3% NaCl		−9.38	2	29
10Ni–Cr–Mo–V	3% NaCl		−9.24	2	29
10Ni–Cr–Mo–V	3% NaCl		−9.11	2	29
12Ni–5Cr–3Mo					
	Air		−9.49	2.56	
4340	3% NaCl		−9.10	2.56	29
	3% NaCl		−8.52	2.56	
	Air		−9.58	2	
	3% NaCl, 10 Hz		−9.44	2	
2Ni–5Cr–5Mo	3% NaCl, 1 Hz		−9.39	2	29
(maraging steel)	3% NaCl, 0.1 Hz		−9.08	2	
	3% NaCl, 1 Hz				
	square wave		−9.59	2	
AISI 4145	Air 15 Hz		−10.96	2.8	
Cr–Mo steel	0.1 N H_2SO_4 1.7 Hz		−9.51	2.02	120
	0.7 Hz		−8.74	1.7	
	Dry argon, 20 Hz		−9.82	2	
	Water vapour 10 Hz		−9.70	2	
	4 Hz		−9.56	2	
AISI 4340	2 Hz		−9.35	2	112
	1 Hz		−9.12	2	
	0.5 Hz		−9.0	2	
	0.1 Hz		−8.35	2	
	Air		−11.86	3.31	
3.5% Ni–Cr–Mo–V	H_2, 5 psi, $R = 0$		−11.42	3.18	121
	H_2, 5 psi, $R = 0.4$		−12.41	3.94	

TABLE 4
PARIS CONSTANT AND PARIS EXPONENT VALUES FOR STAINLESS ALLOYS

Alloy	Environment	$log_{10} C$	n	Ref.
17-4PH VM	Air	−10.06	2.2	
	Seawater, −0.3 V Ag/AgCl	− 9.2	2.0	
	−0.65 V Ag/AgCl	− 9.2	2.0	
17-4PH AOM	Air	−10.22	2.4	
H-1050	Seawater, −0.2 V Ag/AgCl	−9.59	2.1	111
	−0.65 V Ag/AgCl	− 8.46	1.6	
17-4PH AOM	Air	−10.49	2.3	
H-1150	Seawater, −0.2 V Ag/AgCl	−9.98	2.2	
	−0.6 V Ag/AgCl	−10.49	2.0	
13%Cr–8%Ni	Air	−9.74	2.64	122
20PH	Salt water	−7.6	1.67	
ZOCD 26.1	Air	−14.13	4.5	
	3% NaCl, fcp	−16.81	6.64	
23-CNDU	Air	−10.93	2.93	123
	3% NaCl, fcp, 20 Hz	−12.41	3.97	
	3% NaCl, fcp, 0.5 Hz	−14.39	5.38	
VK A271	Air	−11.00	2.88	124
	Whitewater, fcp	−10.28	2.39	
Uranus 50	Air	−11.96	3.26	124
	Whitewater, fcp	−11.96	3.26	
Inconel 718 (forged)	Air	−17.00	6.00	
Inconel 718 (forged)	5 000 psi H_2	—	—	
Inconel 718 (cast)	Air	−20.55	7.55	
Inconel 718 (cast)	5 000 psi H_2	−18.28	7.17	125
Waspalloy	Air	−21.74	8.09	
Waspalloy	5 000 psi H_2	−15.26	6.13	
Alloy 903	Air	−14.49	9.79	
Alloy 903	5 000 psi H_2	−17.00	6.55	

crack growth data and the lower bound of all of the data points fits reasonably well to the straight line given by

$$\frac{da}{dN} = [1.5 \times 10^{-6} e^{-4n}(K)^n] \text{ m/cycle} \tag{2}$$

This equation implies that the separate Paris law sections of curves relating to points lying on the lower bound line (eqn (2)) have different slopes but a common pivot point at $\Delta K = 54.6$ MN m$^{-3/2}$ and

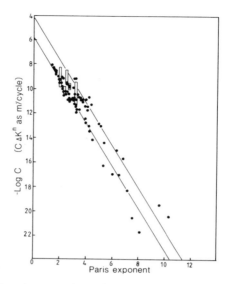

FIG. 12. Correlation between $\log_{10} C$ in Paris equation and exponent n for various air fatigue and corrosion fatigue data.

$da/dN = 1.5 \times 10^{-6}$ m/cycle. The data plotted in Fig. 13 include both air data and corrosion fatigue data. In some cases, the presence of the environment causes the da/dN versus ΔK plot for tests in air to move parallel to itself to larger da/dN at a given ΔK, as illustrated by the vertical columns in Fig. 12. In other cases, corrosion fatigue increases da/dN more at low ΔK than at high ΔK so the net result is a reduction of the exponent n in the Paris equation.

This intermediate ΔK range where the Paris law is approximated is a useful region to consider for investigations into the mechanism of corrosion fatigue because it is reasonably certain that in this range, growth of both air and corrosion fatigue is often by a ductile striation mechanism. Tomkins and Biggs[51] proposed a mechanism for ductile striation formation that leads to a maximum growth rate per cycle of half the crack tip opening displacement. Recently, Krasowsky and Stepanenko[52] have proposed a slightly different interpretation but the Tomkins model is essentially accepted for air fatigue growth by ductile striation formation. However, all fatigue growth is not by ductile striation formation, Forsyth et al.[53-55] having reported the occurrence of brittle striations in an aluminium alloy. Wanhill[56] has examined brittle

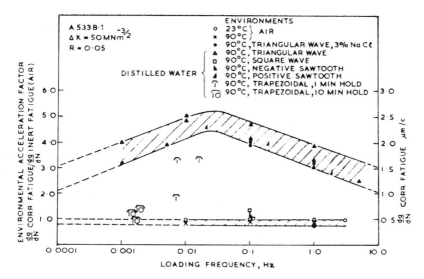

FIG. 13. Summary of A533B corrosion fatigue data in terms of cyclic frequency, from Atkinson and Lindley, Ref. 58.

striation markings in aluminium alloys, mild steel and Ti–2.5%Cu and concludes that they result from limited cleavage in the material ahead of the crack tip that is influenced by the environment, but this cleavage becomes arrested when it runs into material not influenced by absorption. Although not strictly corrosion fatigue data, the work of Ritchie and Knott[57] on the air fatigue of temper embrittled steels is also relevant. They examined the effect of R ratio on fatigue crack growth and showed that when increasing R increased da/dN the fracture surface contained intergranular facets or other 'static' modes of failure in addition to the ductile striations due to the alternating stress.

The point about these various observations is that there may exist several possible mechanisms during corrosion fatigue crack growth and the overall growth rate will be determined by the relative kinetics of the alternative processes. Thus, it is unlikely that a general explanation for corrosion fatigue will evolve and, as with stress corrosion, particular mechanisms may be relevant to particular systems or, alternatively, more than one mechanism may contribute to crack growth.

To attempt to elucidate possible mechanisms of corrosion fatigue it is necessary to concentrate attention on the crack tip, a region of rapidly

deforming material where new surface is being created at a rate that cannot be easily defined. However, it could be argued that the effective gauge length at the crack tip is the crack opening displacement, δ, so the relevant strain rate for the tip could be calculated on that basis.
Taking

$$\delta = \frac{\alpha K^2}{\sigma_{yield} E}$$

where E is Young's modulus, σ_{yield} the yield strength and α is a constant, then the strain rate, $\dot{\varepsilon}$, will be given by

$$\dot{\varepsilon} = \frac{d \ln}{dt} \left(\frac{\delta_t}{\delta_0} \right) = \frac{1}{K^2} \frac{dK^2}{dt}$$

where δ_t and δ_0 refer to the crack opening displacement at time t and zero respectively. If the stress is applied as a function of time, i.e. $\sigma = f(t)$, then

$$\dot{\varepsilon} = \frac{1}{(f(t))^2} \frac{d(f(t))^2}{dt} = \frac{2f^1(t)}{f(t)}$$

For monotonic loading, the strain rate at the crack tip will be very high initially, because the effective gauge length is very small, but will decrease with increasing load as the gauge length increases. Similarly, triangular wave forms give an initially high $\dot{\varepsilon}$ that decreases to the maximum load. What happens during unloading depends upon the details of reversed plastic flow, crack closure, crack extension and crack tip profile. For sinusoidal loading the crack tip strain rate will be zero at the minimum stress, reach a maximum during the loading cycle but fall to zero again at the maximum stress. Thus, in a cyclic loading situation, the region at the crack tip experiences high strain rates and there is the continual generation of new surface if the crack is growing. Hence the presence of an aqueous environment can influence the growth either by enhanced anodic dissolution at the newly created surface, or by passivating reactions or by absorption of reaction products.

If data on the effects of cyclic frequency and load wave form are examined, it is clear that corrosion fatigue is enhanced at low frequency.[58] This presumably means that the corrosive environment is in contact with the tip long enough to make some contribution to crack growth by one or more of the above mechanisms. For instance, the data of Atkinson and Lindley[58] show a maximum enhancement of about × 5

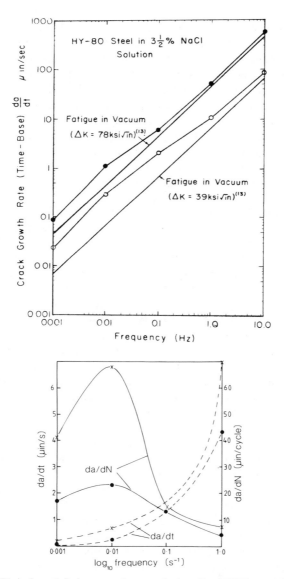

FIG. 14. (a) Time based fatigue crack growth data for HY80 steel from Ref. 59. Loading (ksi $\sqrt{\text{in}}$): ○, $38 < \Delta k < 40$; ●, $76 < \Delta k < 80$. (b) Comparison of the time based fatigue crack growth rate data and cyclic crack growth rate data presented in (a). Loading (ksi $\sqrt{\text{in}}$): ×, $\Delta \times = 78$; ●, $\Delta K = 39$.

at a frequency of ∼0.03 Hz for A533B steel at 90 °C in distilled water (Fig. 13). Similarly, data on HY80 in 3.5% NaCl[59] plotted as log da/dt versus log frequency also show a peak effect at about 0.01 Hz, (Fig. 14(a)). If these latter data are replotted both in terms of da/dN and da/dt versus log frequency an important fact is clarified (Fig. 14(b)). Whereas the enhanced growth rate per cycle is maximized at about 0.01 Hz, the rate of additional growth of the crack due to the environment actually increases progressively with frequency. Thus, the corrosive environment is influencing growth rate at high frequency cycling, but the time of action is so short and the base growth rate due to the mechanical action of the cyclic stress so high that the additional effect of the environment seems negligible, especially when plotted on a log–log scale.

The data drawn in Fig. 13 also indicate that wave form is important and correlate with the experimental evidence in existence showing that crack extension occurs in the loading half cycle.[60] Barsom's data[61] have also shown a similar dependence on load wave form though in the discussion to that paper Hudak and Wei present data for 7075-T651 aluminium alloy apparently showing no effect of wave form. However, when an effect of wave form does exist, square wave and negative saw tooth (i.e. very short rise times) have little or no effect. The maximum effect occurs with sinusoidal, triangular and positive saw tooth wave forms. As indicated in Fig. 13, hold times at peak load tend to lessen the enhancement due to corrosion.

On the basis of crack extension enhancement due to the environment occurring mainly in the loading half cycle, the nil effect of square wave and negative saw tooth is easy to understand, because the environment has virtually no time to contribute because of the very short rise time. The effect of hold time is more difficult to rationalize. There is some uncertainty as to how to calculate frequency in trapezoidal wave form load holding experiments and it could be argued that the equivalent frequency should be plotted as 1/(rise time) because that relates to the period when new surface is being created in the loading half cycle, although even that point is not universally accepted. The hold time data in Fig. 14 refer to rise times of about 7 s and 80 s, i.e. equivalent to 0.07 Hz and 0.006 Hz, but appropriate adjustments to the plotted data in Fig. 13 make no difference to the clear indication that holding at maximum load reduces crack enhancement significantly. The reason for this could lie in the necessity for continued plastic deformation at the crack tip if the active environment is to produce corrosion fatigue. Experimental evidence can be quoted in support of this hypothesis. With systems showing corrosion fatigue but not stress corrosion, slow monotonic loading to fracture of precracked specimens can sometimes be accomplished in the environment without any measurable growth of the air fatigue precrack. For instance, Denton[62] has shown that loading precracked specimens of HY80 in seawater at cross head speeds in the range of 10^{-5} to 10^{-6} in/s and at an applied potential of -950 mV sce gave no crack growth prior to the onset of fast unstable fracture. Thus, just a deforming crack tip plastic zone and the presence of a suitable environment may not be sufficient to cause crack growth. Rather, enough time of contact when the crack tip is experiencing a critical strain rate may be the essential factor.

The situation may well be different for systems that display stress corrosion and involving cyclic loading where, if K_{max} is greater than K_{ISCC}, stress corrosion can be expected to produce crack growth throughout the hold time. In their later paper, Atkinson and Lindley[63] make the point that rise time is the important parameter rather than total cycle time for K_{max} less than K_{ISCC}. They present a schematic model predicting reduced corrosion fatigue with increasing hold time for below K_{ISCC} data but enhanced crack growth rates with increasing hold time for above K_{ISCC} data.

Another fact that supports the contention that it is necessary for newly created surface at the crack tip to be in contact with the active environment is that corrosion fatigue data obtained from precracked specimens sometimes show transients in crack growth at the onset of a test programme. Often, a higher than expected growth rate is experienced initially which slows down to a steady state corrosion fatigue growth rate for the particular applied ΔK. This suggests that an incubation period is necessary to establish quasi steady state fatigue conditions, e.g. a stable crack tip chemistry is necessary that depends on the range of strain rates involved. Thus, hold times may allow micro-creep to occur that affects the strain in the next cycle of a trapezoidal loading wave form and thereby influences the corrosion fatigue rate because both crack tip strain rate and crack tip chemistry are changed by the hold time.

5. CRACK GROWTH AT LOW STRESS INTENSITY FACTOR AMPLITUDES

As was indicated in Fig. 11, many fatigue and corrosion fatigue data show a large dependence of da/dN on ΔK at low ΔK values, and there may exist a threshold stress intensity, ΔK_{th}, below which fatigue crack growth will not occur. The evaluation of a threshold stress intensity is lengthy and difficult experimentally because the crack growth rates are so small. For instance, Vosikovsky[64] has measured crack growth rates of only about 7×10^{-10} m/cycle for an X65 steel in seawater and Irving and Beevers[65] a rate of 2.5×10^{-10} m/cycle for a Ti-6Al-4V alloy *in vacuo*. Such growth rates are of the order of an interatomic spacing per cycle. Typically, the ΔK at threshold could lie in the range of 5 to 10 $MN/m^{-3/2}$ which corresponds to a cyclic crack opening displacement of only about 3×10^{-8} μm for steel,[66] calculated from

$$\delta \cong \frac{0.11 \, (\Delta K)^2}{E\sigma_{yield}} \tag{3}$$

so the observed crack growth rates are very much less than the cyclic crack opening displacement at ΔK values near the threshold stress intensity amplitude.

At low ΔK values, striations are often not observed on the fracture surfaces in either air fatigue or corrosion fatigue. Thus, it seems probable that very low crack growth rates are achieved by small increments in crack length along isolated sections of the crack front. This is quite feasible because the fracture surfaces generated by fatigue and corrosion fatigue are far from planar when examined at high magnification and isolated segments of crack advancement that occurred to give an overall very low crack growth rate is a realistic model.

Thus, the evaluation of ΔK_{th} values under corrosion fatigue conditions can involve very long test times. For instance, a growth rate of 2.5×10^{-10} m/cycle at a frequency of 0.01 Hz requires 12.5 years for a millimetre of crack growth. Worthwhile data can be produced with less crack growth and a higher test frequency but the point being made is that determining ΔK_{th} values is a tedious, time consuming affair and the low ΔK_{th} values that result combined with realistic limits of non-destructive testing for cracks in real structures makes the relevance of ΔK_{th} values, if they exist, debatable. As indicated in Fig. 11, da/dN increases rapidly with ΔK near ΔK_{th} which suggests that in this region the growth per cycle is not simply due to the increased surface area exposed per cycle caused by the crack opening displacement because that would lead to a $(\Delta K)^2$ power law. Also, the increments per cycle are orders of magnitude less than $\Delta \delta$ (eqn (3)), so it must be assumed that either significant reverse plastic flow occurs at low ΔK, that the crack opening displacement equation is not valid at low ΔK or that the mechanisms involved in the very low crack growth rates are not related to the magnitude of the crack opening displacement at all.

It should be noted that for $R = 0$, the crack tip plastic zone size $2r_y$ will be about $1/3$ $(\Delta K/\sigma_{\text{yield}})^2$, i.e. $\cong 50$ μm for a typical steel with $\Delta K = K_{\text{max}} = 5$ MN/m$^{-3/2}$. Hence, the plastic zone size and crack opening displacement are both relatively large compared with the crack growth per cycle. Thus, if the presence of the environment causes an effect associated either with the COD (say oxidation of the surface) or with the plastic zone size (say hydrogen absorption), the increase in crack growth rate due to the environment might be large. Newmann et al.[67] have shown that for copper tested in air, crack growth was by Stage I cracking for growth rates less than 10 nm/cycle but by the Stage II mechanism for growth rates greater than 30 nm/cycle. The mechanism could be altered at will by changing the ΔK level. In vacuo, however, no Stage I crack propagation was found even for da/dN as low as 0.1 nm/cycle. For Fe–3% Si, Newmann also showed a transition from ductile to brittle striations when the crack opening displacement rate was increased from 8 μm/s to 80 μm/s. The brittle striated surface was on a (100) plane but because of the absence of electron chanelling patterns these were considered to be associated with more than 10% plastic strain rather than true cleavage surfaces.

The general view then is that both Stage I and brittle striations involve microplasticity mechanisms to achieve growth and the influence of the environment may be through an absorption or oxidation mechanism.[68] The effect of the environment will usually be to enhance crack growth but not invariably so. For instance, preventing crack resharpening during compression[69] or the encouragement of crack branching[70] can reduce the growth rates.

Generally, both the presence of an environment and increasing the R ratio decrease ΔK_{th}, possibly sometimes for different mechanistic reasons or possibly because in each case an additional cracking mechanism sensitive to the maximum tensile stress rather than the ΔK comes into operation.

Returning to the concept that the crack growth/cycle in the absence of environmental effects is dependent as a maximum on the cyclic crack opening displacement[66] ($\Delta\delta = 0.11$ $(\Delta K)^2/E\sigma_{\text{yield}}$), which is independent of R, a dependence on R can be introduced via crack closure. If $K_{\text{eff}} = (0.5 + 0.4R)\Delta K_{\text{apparent}}$, then $\Delta\delta$ becomes equal to 0.11 $(0.5 + 0.4R)^2\Delta K^2/E\sigma$ yield or some similar R dependent function and $\Delta\delta$ at a given ΔK will increase with increasing R. For the range of R from 0 to 0.9 say, $\Delta\delta$ increases by a factor of 3. This is not sufficient to account for measured R effects in either air or aqueous corrosion fatigue conditions where, if an R effect exists, it often induces orders of magnitude changes in crack growth rate. Tomkins's[66] data showed striation spacings very dependent on R, increasing significantly with increasing R and obeying a ΔK^2 power relationship, but the striation spacings implied crack growth rates much greater than those observed for A533B. The measured growth rates for A533B were significantly less than would be predicted by the striation spacing at low ΔK values for the air tests, but the agreement was much better for the data obtained in simulated PWR water.

The facts that emerge most strongly from the literature are that whereas it is clear that threshold stress intensity amplitudes exist, in the sense that below a particular ΔK for a system the crack growth rate will often approach zero and be at least as low as an atomic spacing per cycle, and that in some cases the crack growth rate is significantly influenced by the R ratio, the detailed reasons for the effects are uncertain. Bearing in mind that corrosion fatigue enhancement appears likely to be greatest with cyclic loading frequencies of about 0.01 Hz there seems no alternative but to grasp the nettle and proceed with longterm low ΔK, low frequency test programmes to try to understand the underlying effects.

In essence, the aim of research into corrosion fatigue is to supply engineers with adequate information for them to make design decisions. In principle, they must relate operational stress transients to material–environment properties so that fatigue cracks do not grow to unacceptable dimensions within the design life. The decisions may be based on either da/dN data or on $S-N$ curve data. For the latter, some materials show a fatigue limit. However, it is well known that the existence of a fatigue limit does not necessarily imply that cracks have not been initiated by the alternating stresses but rather that any initiated cracks arrest and remain non-propagating. Similarly, non-propagating cracks can be initiated below the limiting stress for total failure by stress corrosion cracking of plain specimens,[71] the criterion for non-propagation being related to the crack tip stress intensity factor because of the influence of the latter upon the time dependence of the creep rate after initial loading. Also, non-propagating cracks in air tests on a martensitic–ferritic steel[71,72] were shown to be influenced by microstructural factors via some fracture mechanics control although the exact K values may be uncertain because the validity of the calculation is uncertain for these small cracks.[73] As well as research that has provided actual fatigue data from crack growth rate studies or times to failure, there have been attempts to predict corrosion fatigue crack growth data from mechanics principles[74] and also to account for threshold and R effects. Krafft has argued that crack extension can be likened to the attainment of tensile instability for a ligament of material at the crack tip. He calls the width of material experiencing the equivalent of the ultimate tensile stress the process zone, d_T, and by measuring the tensile and cyclic loading properties of the material Krafft establishes parametric curves that depend on loading rate, strain hardening rate and stress relaxation rate. These can be compared with fatigue data to evaluate d_T for a particular material. When this has been accomplished, the effects of frequency and R ratio can be reasonably predicted for many materials. The extension of these ideas to corrosion fatigue involved the addition of the corrosive attack that occurs at the crack tip. Briefly, Krafft's approach takes cognizance of the importance of microstructural factors and crack tip strain rate in the detailed crack extension process but relates them to a single process zone dimension, d_T, through which limited experimental data can be used in a predictive manner.

6. MODELS AND MECHANISMS FOR CORROSION FATIGUE CRACK PROPAGATION

As previously described in relation to Fig. 11(c), (d), (e), there are three types of log da/dN versus log ΔK plot that may result when a material is fatigue loaded in the presence of an aqueous environment.

Figure 11(c) is typical of the behaviour exhibited by systems for which stress corrosion cracking does not occur under static loads and has been termed 'true corrosion fatigue' (Austin and Walker).[75] As can be seen, the environmental fatigue curve exhibits different values of C and n in the Paris equation from those relevant to an inert environment. In contrast, Fig. 11(d), (e) shows behaviour typical of systems which exhibit stress corrosion cracking under static loads when the stress intensity exceeds a threshold value designated K_{ISCC}. For the system shown in Fig. 11(d) environmental enhancement of cyclic growth rates is only exhibited when the stress intensity range ΔK is such that the maximum stress intensity attained during the fatigue cycle exceeds K_{ISCC}. This has been termed stress corrosion fatigue. (There are some potential difficulties associated with such expressions since, fractographically and in other ways, these modes of failure can sometimes be indistinguishable. It should not be assumed therefore that the expression 'stress corrosion fatigue' implies a different mechanism of failure from that which operates without cyclic loading.) For the system shown in Fig. 11(e), stress corrosion fatigue appears to be superimposed on true corrosion fatigue so that environmental enhancement of cyclic growth rates is also exhibited when $K_{max} < K_{ISCC}$. The latter type of behaviour is the most common type of behaviour shown by material/environment systems exhibiting so called 'stress corrosion fatigue'.

A simple explanation of the type of system shown in Fig. 11(d) was formulated by Wei and Landes in their linear summation model of fatigue crack growth. Wei and Landes proposed[76] that the corrosion fatigue propagation rate $(da/dN)_{cf}$ is equal to the sum of the individual contributions of the environmental rate of attack $(da/dN)_{scc}$ and of the cyclic-dependent mechanical fatigue crack propagation rate $(da/dN)_F$, i.e.

$$\left(\frac{da}{dN}\right)_{cf} = \left(\frac{da}{dN}\right)_{scc} + \left(\frac{da}{dN}\right)_F \tag{4}$$

On a time base eqn (4) reduces to

$$\left(\frac{da}{dt}\right)_{cf} = \left(\frac{da}{dt}\right)_{scc} + f\left(\frac{da}{dN}\right)_F$$

where f is the cyclic frequency and the implication from this latter equation is that as the frequency decreases the environmental time-dependent component should increase.

It is clear that the linear summation hypothesis cannot satisfactorily account for the phenomenon of true corrosion fatigue since in systems exhibiting such behaviour no stress corrosion cracking is observed so that $(da/dN)_{scc} = 0$ in eqn (4) and corrosion fatigue crack growth rates would be predicted to be identical to those in an inert environment. However, it has been shown[75] for certain metal–environment systems that the corrosion fatigue behaviour shown in Fig. 11(c) can be transformed to that typified by Fig. 11(e) and vice versa by altering the mechanical parameters, stress rates and frequency. Thus, reducing the stress ratio and/or increasing the frequency suppresses the stress corrosion plateau shown in Fig. 11(e) and the activation reverts to that shown in Fig. 11(c). This implies that for a given stress ratio a characteristic frequency F^* exists or at a given frequency a characteristic stress ratio, R^*, exists at the interface between the two kinds of behaviour. Austin and Walker[75] proposed a 'Process Competition Model' to enable quantitative assessment of corrosion fatigue crack growth behaviour and the evaluation of F^* and R^*. Unlike the Wei and Landes model,[76] which assumes that the processes of stress corrosion and fatigue (or true corrosion fatigue) are additive, the Process Competition Model assumes these processes are mutually competitive. Consequently, on this basis, the crack will propagate by the fastest available mechanism pertinent to the applied stress intensity.

The interface between true corrosion fatigue and stress corrosion fatigue can be expressed as

$$\left(\frac{da}{dt}\right)_p \frac{1}{F} = c[K_p(1-R)]^n$$

where K_p = stress intensity at onset of stress corrosion plateau and $(da/dt)_p$ = plateau growth rate. Consequently,

$$F^* = A_{(1-R)}n \text{ and } R^* = 1 - \left(\frac{A}{F}\right)^{1/n}.$$

where

$$A = \left(\frac{da}{dt}\right)_p \cdot \frac{1}{c(K_p)^n}$$

Loading wave form can be allowed for by assuming that the stress corrosion fatigue crack growth rate over the relevant stress intensity amplitude is the average stress corrosion rate rather than the stress corrosion rate at the maximum stress intensity during the cycle.

The phenomenon of true corrosion fatigue behaviour therefore seems to be due to the synergistic action of localized corrosion reactions at the crack tip and cyclic loading rather than simply a static environmental crack growth contribution superimposed on a cyclic fatigue crack growth mode. The detailed manner in which the localized corrosion reactions at the crack tip and cyclic loading interact to produce corrosion fatigue is at present the subject of much debate in the open literature. It is clear that any detailed mechanism of environmental crack growth must be able to explain all of the experimental facts particularly in relation to the effects of cyclic frequency, loading wave form, mean stress (i.e. stress ratio), applied potential and the pH of the environment (especially the pH of the environment near the crack tip). While it is primarily the potential of the metal and the pH of the environment within the crack enclave that will determine which of the various electrochemical reactions are possible at the crack tip, mechanical variables such as frequency, wave form and mean stress may influence reaction kinetics and in doing so determine which reactions actually occur at the crack tip.

For example, though the potential of the metal at the crack tip and the pH of the solution within the crack enclave may indicate that two electrochemical reactions are thermodynamically possible, the presence of dynamically strained material at the crack tip and the generation of slip lines intersecting the metal surface in contact with the environment may result in one of the two reactions being more highly favoured on the bare metal so produced. Of the various electrochemical reactions that are possible on the surface of metals exposed to a corrosive environment the two that are of greatest interest in producing enhanced crack growth rates are (i) metal dissolution which can be represented in general terms for any metal, M, by the equation $M \rightarrow M^{n+} + ne$ and, (ii) the cathodic deposition of hydrogen which by adsorption and absorption causes embrittlement of the metal. The latter can be represented by the equation, $H^+ + e \rightarrow H_{ads} \rightarrow H_{abs}$.

7. DISSOLUTION MECHANISMS

Two mechanisms of crack propagation by dissolution have been suggested for environmental cracking due to static loads depending upon the presence or otherwise of structural features (e.g. segregate or precipitate) usually at grain boundaries which cause a local galvanic cell to be set up. In this context the precipitate or segregate may act as the anode in a local cell or by acting as an efficient cathode may cause attack to be localized in the immediately adjacent matrix material. The function of the applied stress in such a system is presumably to prevent the dissolution reaction from becoming stifled by film formation. When pre-existing active paths are non-existent, or are inoperative because of environmental effects, the stress may act to disrupt a protective film formed on the metal surface to expose bare metal and so maintain relatively high localized anodic current densities at the exposed surface.

Crack propagation by dissolution mechanisms involves liquid diffusion of either solvating water molecules or aggressive anions down the crack length to the site of rupture of the protective oxide at planes of slip step emergence at the crack tip followed by dissolution of the newly created metal surface. The rate of crack propagation, V, at a given crack tip potential can be related[77] to the bare surface dissolution current density, i_a, under activation or water diffusion control, to the passivation rate parameter, β, which appears in the current decay formula, $i = i_a \exp^{-\beta(t^* - td)}$ and to t^*, the periodicity of oxide rupture events (td = incubation period for oxide growth).

When $t^* \gg td$ the crack growth rate is given by

$$V = \frac{Mi_a}{Z\ F\rho t^*} \cdot -(1 - \exp(-t^*))$$ (5)

where M = atomic weight of the dissolving metal of density ρ. Z = charge on solvated metal cation and F = Faraday constant.
When $t^* \leq td$ the latter equation reduces to

$$V = \frac{Mi_a}{ZF\rho}$$ (6)

which defines the maximum theoretical propagation rate consistent with experimentally determined potential–current density plots for the metal or alloy in a given environment.

The implication from eqn (5) is that if the creep rate (i.e. strain rate) at the crack tip is insufficient to maintain a bare surface condition, i.e. $t^* \gg td$, then crack propagation rates will be below that given by eqn (6). Conversely, when the crack tip strain rate is too high, mechanical crack advancement may dominate due to ductile rupture at the crack tip. At some intermediate value of crack tip strain rate optimum conditions for environmental enhanced growth will exist. The above quantitative interpretation of stress corrosion cracking can be extended to explain the observed crack propagation rates in true corrosion fatigue where oxide rupture rates and liquid diffusion rates are enhanced by dynamic loading, as discussed in more detail below.

8. HYDROGEN EMBRITTLEMENT MECHANISMS

Adsorped hydrogen may embrittle materials in one of two ways:

1. Directly by adsorping at the crack tip and lowering the surface energy of the material, γ, and thereby lowering the fracture stress of the material, σ_{FR}, according to

$$\sigma_{FR} = \left(\frac{E\gamma}{b}\right)^{\frac{1}{2}}$$ (7)

   where b = interatomic spacing and E = Young's modulus of elasticity.
2. Indirectly by diffusing to some region in advance of the crack tip where the stress or strain conditions are particularly favourable for the nucleation of cracks, e.g. at regions of maximum stress tri-

axiality usually considered to be located at the elastic plastic interface of the plastic zone formed at the crack tip.

However, more important than the detailed mechanism of hydrogen embrittlement is whether environmentally assisted crack growth in aqueous environments is principally assisted by metal dissolution (i.e. corrosion) or by hydrogen embrittlement. In order to understand which of these environmental factors is mainly responsible for the enhanced crack growth in corrosion fatigue, it is necessary to recognize that the electrochemical conditions within the confines of a propagating crack may be significantly different from that at the surfaces exposed to the bulk environment.

9. THE IMPORTANCE OF CONDITIONS WITHIN THE CRACK ENCLAVE

It is conceivable that the electrode potential distribution within a crack enclave may vary with its depth and that the composition of the solution within the crack may be very different from that of the bulk solution. This may, for example, explain why the initiation stage of cracking is not necessarily circumvented, when precracks are introduced into stress corrosion or corrosion fatigue specimens. Initiation may involve the establishment of certain electrochemical conditions within the crack enclave appreciably different from those of the external surfaces.

9.1 Potential Drops Down Cracks

In stress corrosion tests under controlled applied potential, the relevance of the measurement of the potential of the specimen on the external surface where a crack emerges in relation to the potential at the crack tip has been the subject of much discussion in recent years. Vermilyea and Tedman[78] have analysed the situation by assuming inert crack sides and an active (corroding) tip, while Ateya and Pickering[79] have similarly investigated the case of a crack subjected to cathodic charging with hydrogen in acidic environments with a current drain to the crack sides. Both of these approaches showed that the potential drops would be of the order of 100 mV along cracks up to 1 cm in length but with even larger potential excursions occurring when hydrogen bubbles were trapped in the crack. More recently Doig and Flewitt[80] have carried out a similar analysis which predicts even greater potential drops down cracks. The problem with the approaches mentioned above is that while they appear to be applicable in general terms to all systems exhibiting stress corrosion cracking, they do not alway conform to experimental facts. For example, Parkins[81] has shown that, for a C–Mn steel immersed in CO_3–HCO_3 solution, plain specimens having a cylindrical gauge length show essentially the same potential ranges for cracking as precracked specimens having a 1.1 cm long precrack with the probe placed at the mouth of the precrack (Fig. 15). If the potential drop along a crack is relatively large then the potential range for cracking of initially plain specimens would be expected to be appreciably different from that for specimens prepared with narrow precracks. As Fig. 15 indicates, the maximum displacement of the two curves is at most only 20 mV so for this particular system significant potential drops do not exist along the length of the crack. It should be noted that in the system investigated by Ateya and Pickering[79] the potential at the crack tip was observed to approach

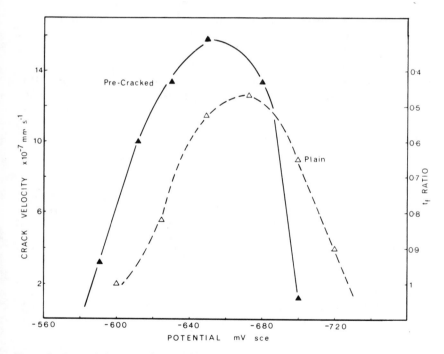

FIG. 15. Potential ranges for cracking of precracked and plain specimens of C–Mn steel subjected to low strain rate tests in CO_3–HCO_3 solution at 75 °C. Cracking susceptibility measured in precracked specimen tests as crack velocity and in plain specimens as time to failure ratio. From Parkins, Ref. 81.

the free corrosion potential of the metal, whereas this could not occur for C–Mn steel in CO_3–HCO_3 solution (Fig. 15) because the free corrosion potential for the latter system is about 400 mV anodic to the middle of the cracking range and at a potential that does not promote cracking.

The basic difference between the system described by Ateya and Pickering[79] and the C–Mn system to which Fig. 15 refers is that the net currents flowing in the potentiostatic circuit between specimen and counter electrode are significantly different for the two cases. Thus, for C–Mn steel in CO_3–HCO_3 the *net* currents flowing between specimen and counter electrode in the stress corrosion cracking potential range are usually very small whereas in the Ateya and Pickering[79] system during cathodic charging the net currents flowing between specimen and counter electrode are relatively much higher. The implication is that significant potential drops are only expected when the net currents (anodic or cathodic) flowing in the potentiostatic circuit are high. The latter has recently been confirmed in unpublished work by the authors carried out on a medium–high strength steel (HY80) exposed to natural seawater. In this investigation the potential drop down cracks of up to 13 mm in length was measured for control potentials in the range − 1500 mV sce to 0 mV sce. Table 5 summarizes the results for a 13 mm long crack. It is clear from these results that significant potential drops along the length of the crack were only observed when significant net anodic or cathodic currents were measured in the potentiostatic circuit.

The basic cause of any variation in the potential of a metal or alloy

TABLE 5

POTENTIAL DROP, NET CURRENT AND SPECIMEN POTENTIAL AT THE TIP OF A 13 mm LONG FATIGUE CRACK IN HY 80 STEEL AS A FUNCTION OF APPLIED POTENTIAL AND CRACK LENGTH AFTER 10 MIN EXPOSURE. EXPOSED SURFACE AREA APPROXIMATELY 15 cm²

Bulk applied potential (mV sce)	Net current flowing (mA)	Potential drop (mV)	Specimen potential at crack tip (mV sce)
− 1500	− 335	+ 495	− 1 005
− 1 250	− 300	+ 560	− 690
− 950	− 1.2	+ 269	− 681
− 700	+ 0.1	+ 37	− 663
− 500	+ 1.2	− 162	− 662
− 250	+ 400	− 360	− 610
± 0	+ 420	− 500	− 500

with distance along the length of a crack, crevice or slot introduced into the metal or alloy is the highly resistive path to current flow offered by the solution within the enclave. This highly resistive path to current flow causes the current density on the walls of the enclave to be significantly lower than on surfaces exposed outside the enclave. The variations in current distribution over the surface of a specimen containing a crack or crevice can be demonstrated using an electrical analogue experiment. In the latter, electrically conducting paper is used to model the current flow through the electrolyte between the surface of the specimen and a counter electrode. Figure 16 shows the flow lines representing the current distribution on the surface of a specimen containing a slot with aspect ratio (length/width) of 5 and in which the externally exposed surface area is equal to that within the slot. The current density acting over any area of the surface is represented in this model by the number of current flow lines intersecting unit length of the line representing the exposed surface of the specimen. It is clear from Fig. 16 that the current density on the external surface of the specimen is high and essentially uniform. However, the current density falls off dramatically just inside the mouth of the slot and is negligible at the base of the slot. Since the local potential of a metal with respect to the immediately adjacent solution is a unique function of the current density (in accordance with the appropriate polarization curves for example) then it can be seen that the above model would predict a variation in potential of the metal with respect to the solution along the length of the crack.

In view of the above model, it might be expected that significant potential drops would occur along the length of the crack in a system exhibiting stress corrosion cracking but they are not invariably observed experimentally. The reason for this is that in the model discussed above, no account was taken of the formation of stable passive films on the crack sides and on surfaces external to the crack enclave which are known to form in such circumstances. These passivating films will significantly alter the current density distribution over the metal surface and consequently will also influence the potential distribution along the length of the crack.

This does not imply that potential drops can never exist along the length of cracks but simply that at potentials where crack propagation occurs due to anodic dissolution the formation of stable passive films along the crack sides precludes the existence of large potential drops along the length of the crack. With reference to the appropriate potentiodynamic polarization curves for different metal environment com-

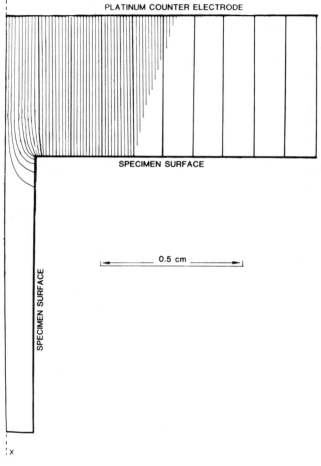

FIG. 16. Electrical analogue of the current flow through the electrolyte between a specimen containing a narrow slot and a platinum counter electrode. XY represents a line of symmetry parallel to the length of the slot and at mid-slot width.

binations, the potentials at which crack propagation due to anodic dissolution is likely to occur can be identified and also those potentials at which significant potential drops along the length of the crack are likely to develop.

Two basic system types which exhibit environmentally assisted crack propagation can be described in terms of their appropriate potentio-dynamic polarization curves. Schematic polarization curves for these two system types are shown in Fig. 17. The curves are determined by varying, at a constant rate, the potential of the metal with respect to a reference electrode and monitoring the corrosion current. The fast sweep rate

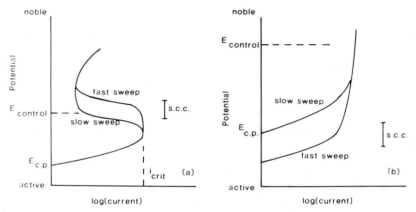

FIG. 17. Schematic potentiodynamic polarization curves for (a) system showing an active → passive transition, (b) a system which does not exhibit an active → inactive transition.

curves correspond to a fast rate of change of the applied potential with respect to time and therefore represents a non-equilibrium situation in which there is insufficient time for the development of the equilibrium film, so the current density remains high. Experience has shown that stress corrosion cracking usually occurs over the range of potentials marked SCC in Fig. 17 in regimes where the relative difference in current (or current density) between the fast and slow potential sweep rate curves is high. The fast sweep rate curve is then indicative of the current density acting locally at the crack tip and the slow sweep rate curve is indicative of the current density acting over the crack sides and surfaces external to the crack enclave. In systems of both types illustrated in Fig. 17, therefore, the stable condition of the metal at the cracking potential is one in which the bulk of the surface of the metal is covered by a passive film.

It is now possible to identify those situations in which a large potential drop is likely to exist along the length of a crack in systems of both types shown in Fig. 17. First consider the situation in a system of the type shown in Fig. 17(a) (i.e. a system which exhibits a so-called active→passive transition). Assume that the specimen contains a pre-crack and that the material is initially oxide free upon exposure to the environment (a situation which would be difficult to realize in practice) so that the free corrosion potential is E_{cp} in Fig. 17(a). Consider what happens when a potential E_c (in the cracking range) is applied to the external surface of the specimen via a potentiostat. In order that the control potential, E_c, is achieved at all points on the exposed surface of the specimen, the current density must rise to a value in excess of i_{crit}. When the potentiostat set to control at E_c is switched on, a high current, I_{max}, will flow between specimen and counter electrode to provide the current density, i_{crit}, necessary for E_c to be achieved on the exposed surfaces external to the crack enclave. Initially, therefore, a current density distribution similar to that shown in Fig. 16 will exist over the surface of the specimen and consequently a potential difference of magnitude approximately E_c-E_{cp} will exist along the length of the crack, if only temporarily. Once i_{crit} is exceeded on the surfaces external to the crack enclave the total current flowing between specimen and counter electrode falls as E_c is approached due to the formation of a stable passive corrosion film on these surfaces. Once the passive film has formed on the external surfaces some current, though small in magnitude, can still flow along the crack enclave. This current acts over very small areas of bare metal close to the interface between the bare metal and previously passivated surfaces inducing locally high current densities sufficient to exceed i_{crit}. The interface will subsequently extend in a progressive manner along the walls of the cracks as indicated in Fig. 18

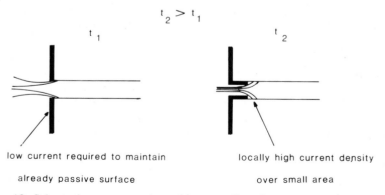

FIG. 18. Schematic representation of current flow in a crack enclave under potentiostatic control illustrating the growth of a passive film.

so that eventually the entire surface within the crack enclave will be passivated and the cracking potential, E_c, will eventually be attained at the crack tip. Conditions conducive to crack advancement at E_c are achieved by the action of the superimposed stress causing plastic deformation at the crack tip which disrupts the passive film covering the surface of the metal in that region thereby exposing bare metal which dissolves at a rate given by the fast sweep rate polarization curve. Because the area of bare metal produced at the crack tip during straining will be small, the current requirement for the dissolution process at that locality can be provided easily by the potentiostatic circuit. For example, for a 1 cm long, 10^{-3} cm wide square-fronted crack in a specimen 1 cm thick the total current required to maintain a crack tip anodic current density of 10 mA cm^{-2} would be 10 μA. In a solution of specific resistance of 25 Ω cm^{-1} the maximum emf needed to drive this current through the solution within the crack would be 2.5 V. The latter can be provided readily by most commercial potentiostats so that cracking conditions could easily be maintained.

The above discussion refers to the situation in which the surfaces of the metal exposed to the environment are assumed to be initially film free. In most practical situations the metal will be covered with an air-formed oxide film and the free corrosion potential of the metal will usually be within the passive range of potentials in Fig. 17(a). This means that in order to achieve the control potential, E_c, lying within the cracking range of potentials, the potential of the metal must move in the active direction over a potential range in which anodic loops in the polarization curve are severely curtailed. Consequently, the air-formed oxide film will have an effect similar to a passive film produced by corrosion reactions, i.e. it will reduce the tendency for large potential drops developing along the length of the crack. Since the air-formed oxide film is present from the outset, the control potential will be achieved very much faster than in the case where the metal is film free.

In a system of the type shown in Fig. 17(b), cracking usually occurs at a potential approximating the free corrosion potential so that under these conditions a negligible potential drop would exist along the length of the crack. If, however, the potential on the surfaces external to the crack enclave were for some reason to be held at a much more noble potential, such as E_c in Fig. 17(b), then the potential drop down the crack would be expected to be significant since the current density distribution over the surface of the specimen, which would be similar to that shown in Fig. 16, could be maintained over a long period of time because a stable passive film does not form over the surface of the specimen. In the limit, the maximum potential drop down the crack will be of magnitude $E_c - E_{cp}$.

So far the effect of cathodic polarization on the development of potential drops down cracks has not been considered. Essentially similar arguments should be applicable to the potential drop existing down the crack as were discussed for anodic polarization. For cathodic polarization, however, the behaviour in systems of both types shown in Fig. 17 should be similar, but being more characteristic of anodic polarization in the system shown in Fig. 17(b). It would be expected, therefore, that the potential of the metal at the crack tip should be at most only a few tens of millivolts removed from the film free – free corrosion potential even though the surfaces of the specimen external to the crack enclave are being controlled at a significantly more active potential where high net cathodic currents are flowing. It is well known that in tests on pre-cracked specimens in which significant cathodic polarization is adopted, the crack growth rates can be over an order of magnitude higher than those observed in specimens tested under freely corroding conditions. This appears to provide evidence in conflict with the previous suggestions. However, hydrogen discharged on the outer surfaces of the specimens is not necessarily immobile and can migrate to the crack tip by surface or sub-surface diffusion so that the material at the crack tip could still be embrittled and exhibit crack propagation rates of the appropriate magnitude. Even so, it might be anticipated that in very thick specimens where diffusional distances for hydrogen would be much longer, hydrogen embrittlement would be more difficult to induce at mid-section than in thinner specimens of the same type. Some results from the literature can be quoted in support of this latter view. Barsom[82] has performed corrosion fatigue experiments on 2 in thick WOL specimens of a 12Ni–5Cr–3Mo steel at an applied potential of – 1200 mV sce and a frequency of 6 cpm. The data are reproduced in Fig. 19 and would appear to indicate that on switching the control potential to – 1200 mV sce from – 800 mV sce there was an increase in crack growth rate at the surface but this was not sustained. The crack growth rate eventually returned to that for corrosion fatigue at – 800 mV sce (the latter approximating the free corrosion potential) and the crack front at the completion of the test was found to be concave towards the mid-thickness. In view of the concave nature of the crack front and the fact that crack length measurements were made on the surface of the specimen by optical means, the validity of the data shown in Fig. 18 is

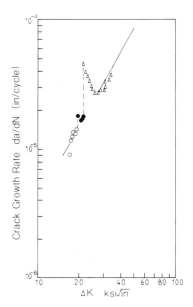

FIG. 19. Corrosion fatigue crack growth data from Barsom, Ref. 82, indicating limited diffusion of hydrogen in thick specimens. ○, 6cpm open circuit; ●, 6cpm 800 mV sce; △, 6cpm 1200 mV sce.

open to some doubt. However, the fact that the crack front was concave towards mid-thickness does imply that crack growth occurred more rapidly at the outer edges of the specimen. This could be explained by the hydrogen generated on the specimen surface by the applied potential of -1200 mV sce causing appreciable localized embrittlement resulting in a marked increase in surface crack growth rate. At mid-thickness, however, the specimens remained unaffected by the imposed applied potential because a significant potential drop along the crack front produced conditions towards mid-thickness which were essentially characteristic of those existing at the free corrosion potential.

In summary, therefore, it seems that whether or not significant potential drops are likely to exist along the length of growing cracks will depend upon whether or not high net currents can be maintained between specimen and counter electrode. Any tendency for the material to form a passivating film precludes the development of large potential drops. It should therefore be possible to predict which control potentials are conducive to the development of large potential drops down the length of cracks or crevices by inspection of the appropriate polarization diagrams for the metal–environment combination.

9.2 Compositional Changes Within the Crack Enclave

When corrosion takes place on metal surfaces exposed to a small volume of solution, as for example in a pit or a crack and where access of fresh electrolyte and dissolved gases is restricted, the composition of the solution within the pit or crack may change markedly from that of the bulk solution. As suggested by Evans,[83] local acidity can develop due to hydrolysis of metal ions in solution according to reactions such as:

$$3Fe^{2+} + 4H_2O \rightarrow Fe_3O_4 + 8H^+ + 2e$$

It can be predicted that hydrolysis will cause the crack tip pH or pH within a pit to equilibrate at around 3.5 when iron is exposed to an aqueous environment at the free corrosion potential. The actual pH developed depends upon the exact nature of the hydrolysis reactions taking place within the crevice.

Thus the reaction

$$Fe^{2+} + H_2O = FeOH^+ + H^+$$

has an equilibrium pH given by

$$pH = 4.75 - \tfrac{1}{2} \log(Fe^{2+})$$

The reaction of ferrous ions to produce ferric ions due to any dissolved oxygen in solution and the consequent hydrolysis of the ferric ions according to

$$Fe^{3+} + 3H_2O \rightarrow Fe(OH)_3 + 3H^+$$

may produce even lower pH values since the equilibrium pH for the latter reaction is given by

$$pH = 1.61 - 1/3 \log(Fe^{3+})$$

Several other reactions are also possible which could also lead to acidification of the solution within the crevice.

Clearly, other concentration changes must also be taking place within the crevice at the same time. In particular the restrictions of limited diffusion must result in an increase in the concentration of cations within the crevice and also, in order to maintain electroneutrality, an increase in the concentration of anions. The indications are that in chloride solutions, for example, the solution may consist of cation and chloride anion concentrations in excess of 1 M.[84]

The significance of both a change in the pH and potential of the metal at the crack tip on possible crack propagation mechanisms can be deduced if the relevant experimental data are plotted on the appropriate potential–pH diagram for the system under investigation. Figure 20,

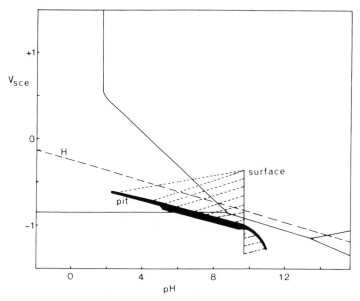

FIG. 20. Influence of cathodic polarization on the electrochemical characteristics of a corrosion pit or crevice (carbon steel in the presence of an aerated solution of NaOH 0.001 M and NaCl 0.001 M) (schematic). From Pourbaix, Ref. 85.

for example, shows the influence of cathodic polarization on the potentials and pH values developed within a simulated pit in a carbon steel immersed in 0.001 M NaCl of pH 10.[85] The extent and directions in which the potential and pH in the simulated pit change are dependent upon the degree of polarization applied to the steel surface external to the pit but the potential within the pit always falls below the line representing the discharge of hydrogen from water.

Similar results have been obtained by Brown[84] in relation to the potentials and pH levels developed in propagating stress corrosion cracks in a high strength steel immersed in a sodium chloride solution (Fig. 21). The fact that the conditions at the crack tip were always favourable for the discharge of hydrogen to occur led Brown to conclude

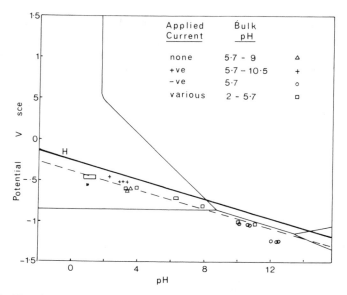

FIG. 21. The electrochemical condition at a propagating crack in AISI 4340 steel. From Brown, Ref. 84.

that stress corrosion cracking in the system investigated was due to hydrogen embrittlement. It is worth while pointing out, however, that under freely corroding conditions dissolution must always be occurring within the crevice and indeed the rate of dissolution will increase with acidity, so that it is also possible under such circumstances for some crack advancement to be due to metal dissolution.

Although the above remarks refer to measurements made in pitting or stress corrosion situations they should also be applicable to corrosion fatigue cracks. The main difference between corrosion fatigue and the other forms of crevice corrosion referred to above is that cyclic crack opening induced by the alternating applied load in corrosion fatigue may result in solution being pumped into and out of the crack producing regular mixing of the crevice solution with the bulk solution. The effect of cyclic loading on solution chemistry modification within corrosion fatigue cracks has recently been considered on purely theoretical grounds by Hartt et al.[86] By considering the volumetric flow rate of the solution exiting from a crack as a function of the crack opening angle, an equation relating the fluid momentum to the other relevant fatigue parameters was deduced. The final equation was:

$$J = \frac{\rho a^3 \theta \omega^2}{4} \int \frac{\cos^2 \omega t}{\sin \omega t + \theta_0/\theta} \, dt$$

where θ = half angular crack opening ω = cyclic frequency
   a = crack length t = time
   ρ = solution density θ_0 = mean crack opening angle

This equation suggests that the solution momentum and therefore the extent of mixing vary directly with 2θ, the crack opening angle and the cube of the crack length. The relationship between the solution momentum and cyclic frequency and mean stress (te θ_0/θ) is more complex. By integrating the above equation between limits $\pi/2\omega$ and $3\pi/2\omega$ (the closing portion of the fatigue cycle) it is possible to express solution momentum as a function of mean stress for a range of frequencies and solution momentum as a function of frequency for a range of mean stress values. Figure 22, for example, shows the relationship between net

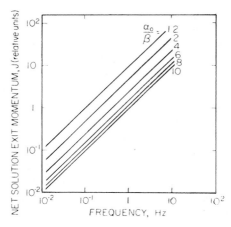

FIG. 22. Net solution exit momentum as function of frequency for various mean stresses. From Hartt et al., Ref. 86.

solution exit momentum as a function of frequency for a range of mean stresses expressed as (θ_0/θ). The results can be summarized as shown in Table 6.

Other factors which are not taken into account in the above equation will also influence the likelihood of crack enclave solution modification such as the relative velocity between the solution and specimen, the

TABLE 6

PROJECTED INFLUENCE OF VARIOUS FATIGUE VARIABLES ON CRACK ELECTROLYTE MIXING WITH BULK SOLUTION

Fatigue variable	Influence on mixing
Crack opening angle, 2θ	Mixing predicted to increase linearly with crack opening angle
Mean stress, θ_0/θ	Mixing predicted to increase linearly with decreasing mean stress and become large when stress function such that crack is closed during part of fatigue cycle
Frequency, ω	Mixing predicted to increase linearly with frequency
Crack length, a	Mixing predicted to increase with cube of crack length
Stress wave form	Related to crack opening angle such that at given frequency mixing predicted to increase the more rapid the rate of crack opening and closing

current density associated with corrosion processes at the crack tip and temperature. The relative velocity between specimen and solution may be important since if this value is low then reingestion of the electrolyte originally present within the crack enclave and expelled during the crack closing cycle may occur. The current density associated with corrosion processes occurring within the crack enclave will also be important since solution modification is more likely to occur for a material subjected to a high rate of corrosion within the fatigue crack than for a material subjected to a low rate of attack. The influence of temperature on solution modification is via its influence on the kinetics of diffusion. Thus, if the momentum of the crack solution expelled from the crack enclave is low or when flow is viscous then any diffusional contribution would be pronounced. Since the rate of diffusion increases with temperature then under the latter circumstances mixing would be enhanced by a temperature rise. The above theoretical approach to crack enclave solution modifications is of little value unless it can be shown to relate to what occurs during actual corrosion fatigue. Clearly, experimental evidence of the occurrence of crack enclave solution modifications and their influence upon crack growth rates during corrosion fatigue is required. No attempts to monitor crack enclave solution compositions appear to have been made so far and it seems, therefore, that if a complete picture of the corrosion fatigue crack propagation process is to be obtained, experiments designed to establish the various aspects of crack enclave solution compositions during corrosion fatigue are desirable.

10. THE INTERACTION OF ENVIRONMENT ASSISTED CRACKING UNDER STATIC AND ALTERNATING STRESS

The effects of loading variables (frequency, wave form and mean stress) on the corrosion fatigue crack growth rates for different materials exposed to various environments appear to show certain distinctive trends depending upon whether or not the stress intensity during the fatigue cycle exceeds K_{ISCC}, i.e. is above or below the stress intensity above which stress corrosion cracking occurs.

10.1 Above K_{ISCC} Behaviour

In general, fatigue crack growth rates on a cycle basis are observed to increase with decreasing frequency. Such effects have been observed for high strength steel[87,88] Al alloys[89,90] and Ti alloys,[91] the results usually being explained by the longer times available for environmental crack growth (stress corrosion) in each cycle as the frequency is reduced. Thus in the limit of testing carried out at very low frequency and with $K_{mean} > K_{ISCC}$ the situation tends towards that of a slow strain rate monotonic loading stress corrosion cracking test. Similarly, the effects of loading wave form in corrosion fatigue systems exhibiting above K_{ISCC} behaviour can be singly related to the different times for which the stress intensity exceeds K_{ISCC}. Thus, wave forms which show significant hold times at stress intensities above K_{ISCC} are expected to increase the time-dependent component of cracking, as previously described. The effect of mean stress (usually expressed in terms of stress ratio, R) would also be expected to be significant in metal environment systems exhibiting above K_{ISCC} behaviour for essentially the same reasons. Thus, as R increases, the time for which the stress intensity will exceed K_{ISCC} during a given cycle will increase so that $(da/dt)_{scc}$ and hence $(da/dt)_{cf}$ should increase.

10.2 Below K_{ISCC} Behaviour

In general, earlier researchers have indicated that cyclic crack growth rates were observed to increase with decreasing frequency in systems fatigued at stress intensities below the threshold value for stress corrosion cracking. Barsom's[45] data for a 12 Ni–5 Cr–3 Mo steel tested in 3% NaCl solution at pH 7 are typical (Fig. 10). Although on a cyclic basis the overall corrosion fatigue crack propagation rates are seen to decrease with increasing frequency, it can be shown that on a time basis the environmental contribution to crack propagation (e.g. da/dt soln $- da/dt$ air or vacuum) increases with frequency. Fatigue crack propagation data for a rotor steel tested in hydrogen gas[92] (Fig. 23) and

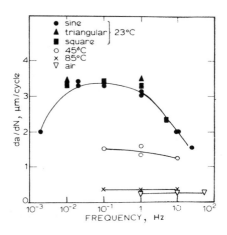

FIG. 23. Effect of frequency and temperature on fatigue crack growth rate for 2Ni–Cr–Mo–V steel in hydrogen. From Smith and Stuart, Ref. 92.

A533B tested in distilled water or 3% NaCl solution[63] (Fig. 13) have revealed that, at least for the systems quoted, a frequency inversion is observed in the cyclic crack propagation rates. There seems to be universal agreement that for systems showing below K_{ISCC} enhancement of crack growth rate, only those wave forms with significant rise times enhance the crack growth rates and that growth rates are independent of the hold time at peak load. As already indicated, the influence of mean stress can be complicated by the possibility of an inherent R ratio dependence of crack growth rates. However, in below K_{ISCC} systems which do not exhibit such intrinsic behaviour it is usually observed that environmental enhancement is essentially unaltered by changes in stress ratio.[46,93]

The observed influence of loading variables on fatigue crack propagation rates on materials showing below K_{ISCC} behaviour indicates quite clearly that in such systems the environmental effect only operates during the tensile loading part of the fatigue cycle. The implication is therefore that it is the loading rate or strain rate at the crack tip which is the most significant factor influencing the crack enhancement process. The situation can be compared with some stress corrosion cracking systems in which the likelihood of cracking appears to be dependent upon existence of a critical strain rate[94] as indicated earlier. Thus, the critical frequency effect illustrated in Figs 13 and 23 could possibly correspond to the critical strain rate producing maximum environmentally assisted cracking susceptibility in certain metal/environment combinations as indicated by Parkins[94] and represented schematically in Fig. 24. While the frequency inversion effect and the effect of loading

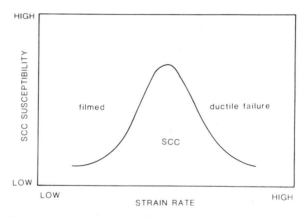

FIG. 24. Schematic representation of dependence of susceptibility to stress corrosion cracking on strain rate (extension rate in constant displacement rate experiments).

wave form on fatigue crack propagation rates below K_{ISCC} appear to be consistent with a mechanism of environmental enhancement dependent upon the rate of oxide rupture and rate of bare metal production they do not enable slip dissolution and hydrogen embrittlement crack advancement processes to be differentiated. Thus, while the rate of dissolution of a metal is clearly dependent upon the generation of bare metal by oxide rupture and slip there is some evidence to suggest that, at least in some systems, hydrogen embrittlement processes also require the preliminary rupturing of an oxide film covering the metal surface. Thus, Scully and Powell[95] consider that the stress corrosion cracking of α-titanium alloys is an example of slow strain rate hydrogen embrittlement induced by the rupturing of an oxide film on the surface of the alloy followed by the cathodic deposition of hydrogen on the bare metal so produced which causes hydride to be nucleated on operative slip planes leading to cleavage.

11. THE INFLUENCE OF POTENTIAL (CATHODIC PROTECTION)

The influence of applied potential, both anodic and cathodic to the free corrosion potential, on the fatigue crack propagation rates of materials exposed to aqueous environments does not appear to have been widely investigated. The vast majority of corrosion fatigue work has been concerned with fatigue crack propagation rates in different environments under freely corroding conditions. Where applied potentials have been employed usually the aim has been to investigate the effect of cathodic protection to reduce fatigue crack propagation rates, particularly in relation to structural steels used in offshore structures.

Cathodic protection is widely employed to reduce material losses due to corrosion in marine environments where it has been estimated that an applied potential 200 to 300 mV more active (negative) than the free corrosion potential will reduce the corrosion rate of carbon steel by 99%.[96]

In view of the fact that some steels may be subject to hydrogen embrittlement cracking and since it can be difficult in practice to maintain uniform potentials over the entire surface of a protected structure so that local potentials may fluctuate around the desired protection potential, the effects of both under- and over-protection are of interest.

There seems to be general agreement that moderate cathodic protection can produce a corrosion fatigue performance of initially smooth specimens of the lower strength structural steels comparable to that in air.[33,97,98] However, in higher strength structural materials which are more prone to hydrogen embrittlement, cathodic over-protection can enhance corrosion fatigue crack initiation at lower cyclic frequencies.[99]

The influence of cathodic protection on fatigue crack growth rates is less conclusive. There is evidence that for high strength steels cathodic polarization may be of little benefit in reducing corrosion fatigue crack growth rates and that over-protection may be decidedly detrimental. As previously indicated, high strength steels are prone to hydrogen embrittlement so that these latter observations are hardly surprising.

However, for API X65 pipeline steel (which is a low strength steel and therefore not usually regarded as being susceptible to hydrogen embrittlement) it has been shown that high cathodic potentials (i.e. over-protection) can increase crack growth rates (in some regimes of ΔK) by as much as 50 times over those observed in air.[64] On the other hand, studies of structural steels in saline environments have indicated that cathodic potentials of about -780 mV sce[100] or protection via zinc anodes[101] can reduce fatigue crack growth rates to those observed in air. However, with cathodically protected materials, potential drops may exist down cracks so the effectiveness of cathodic protection may decrease as the crack front grows into the material from the surface. Barsom has provided some evidence in support.[82] In corrosion fatigue studies on 2 in thick WOL specimens of 12 Ni–5 Cr–3 Mo steel in 3% NaCl solution at cathodic potentials, measurements of the pH of the solution adhering to the fracture surfaces of the specimens after failure indicated that the solution in the vicinity of the crack tip was acid and that the pH increased towards the bulk solution value with increasing distance towards the outer surfaces. From this it was concluded that the potential at the crack tip must have at least attained the free corrosion potential in order for some dissolution and hydrolysis of the resulting corrosion products to occur. The above seems an important point to bear in mind when selecting crack growth specimens for experimental corrosion fatigue studies.

Although from an engineering viewpoint the performance of corrosion fatigue experiments at potentials anodic to the free corrosion potential would appear to be superfluous it is possible that some useful information about the detailed process of environmental crack enhancement (i.e. dissolution or hydrogen embrittlement) could be obtained from such experiments especially if coupled with careful fractography of the corrosion fatigue crack surfaces.

12. THE PREVENTION OF CORROSION FATIGUE

There are two possible approaches to attempting to prevent the occurrence of corrosion fatigue failure in structures. In the first, the engineer's approach, the working stresses to which the material is to be subjected are limited so that the stresses never (or rarely) exceed those corresponding to some corrosion fatigue threshold during the lifetime of the component. In the second, the electrochemist's approach, attempts are made to change the environmental conditions by controlling the applied potential using inhibitors or coating systems so that the damaging corrosion reactions are less likely to occur.

Usually, the protective measures are aimed at preventing corrosion fatigue cracks from initiating rather than stopping cracks from growing once they have initiated. In this respect, most of the successful preventive measures can be regarded as purely surface acting. For example, the fatigue life in a corrosive environment can often be improved by the introduction of compressive stresses into the surface layers of structures by mechanical means such as shot blasting or peening and autofrettage, or by chemical means such as nitriding or carburizing. However, the depth of the cold worked metal developed by such processes is small so that if the layer is subsequently removed, for example by slow general corrosion, susceptibility to cracking may well return.

Possibly the most widespread method of preventing corrosion fatigue crack initiation has already been touched upon in the previous section and that is by cathodic protection. Cathodic protection is a remedial measure that can often raise the fatigue limit of a material exposed to a corrosive environment to the corresponding fatigue limit *in vacuo.* Cathodic protection can be achieved either by coupling the structure to be protected to a metal more base in the electrochemical series or by applying a cathodic potential or current to the structure via an external battery or generator. Examples of the former method of cathodic protection include zinc or cadmium plating of steel. Electrodeposited coatings of zinc and cadmium[102] usually offer slightly better protection than hot dipped or sherardized coatings because the compressive stresses produced by electroplating can in themselves be effective in increasing the fatigue limit as described above. Other metallic coatings on steels such as nickel which may be very efficient in preventing corrosion under static conditions have little or no effect on corrosion fatigue. Under static conditions the nickel layer is effective by providing a barrier between the corrosive environment and the substrate metal. Once the nickel layer is ruptured, due to the application of a tensile stress, attack of the substrate metal can occur unhindered and may even be enhanced due to the galvanic action between nickel acting as cathode and steel acting as anode.

Non-metallic coatings such as epoxy resins, rubber and enamel which do not incorporate additional inhibitors have been shown to be beneficial in increasing fatigue lives in corrosive environments though usually only when accompanied by additional surface treatments to the material before application of the coating. For instance, Gerard and Sutton[103] have shown that significant increases in the endurance of aluminium alloys tested in salt spray could be achieved by coating the alloys with stoved synthetic resins after an initial anodizing heat treatment in a chromic acid bath. The principal danger associated with the use of such non-metallic coatings is that, if the coatings suffer mechanical damage, corrosion of the base metal can occur unhindered.

Paint coatings applied to structural members have also been shown to improve fatigue resistance. Paints are usually of complex chemistry and accomplish protection by a combination of inhibitive action, barrier action and, in the case of base metallic paints, a form of cathodic protection action. For a detailed discussion of the protective mechanism of paint schemes the reader may consult treatises on the subject such as *Principles of Metal Surface Treatment and Protection* by D. R. Gabe[104] or *Protective Painting of Structural Steels* by F. Fancutt and J. C. Hudson.[105]

An alternative form of protection which has found considerable success in preventing stress corrosion cracking is anodic protection. This technique relies on the development of a protective passive film on the surface of the material which can be achieved by raising the potential of the material by the application of anodic current from an external source. Anodic protection has been shown[106] to improve the endurance limit of plain carbon and stainless steels in an oxidizing environment. The improvement is greatest with materials such as stainless steels and titanium alloys which rely on a highly protective oxide for their corrosion resistance.

The preventive methods discussed above are essentially surface active, but it is possible to consider changing the composition or structure of an alloy to improve stress corrosion or corrosion fatigue resistance. This has been the subject of some research but it has usually been found that metallurgical solutions are not economically viable. For example, improvements to stress corrosion cracking resistance of steels and aluminium alloys can be achieved by reducing the strength level of these alloys by heat treatment but these are not cheap solutions to the problem when the alloys were originally developed for optimum strength. Similarly, the cracking resistance of austenitic steels can be improved by large increases in the Ni content of the alloy and the addition of 1–2% of titanium to mild steel can have similar effects, but the large amounts of such alloying additions required to improve markedly cracking resistance considerably increase the cost of the steel. For example, it has been estimated that an addition of only 1% Ti to mild steel approximately doubles the cost of the steel for unit weight.

The remaining alternative to either surface active remedies or alterations of the properties of the material is to modify the environment, possibly by the addition of corrosion inhibitors to the environment. Since inhibitive treatments are usually costly, they are normally applied only in closed systems in which the corrosive environment forms the working fluid, e.g. in boilers or heat exchangers. The nature of the inhibitor employed varies according to the metal and corrosive environment concerned but in general all inhibitors are effective either by reacting with the products of the corrosion reactions occurring on the metal surface or by adsorbing on to the metal surface and thereby forming insoluble protective films. Figure 25 shows the effect of a sodium

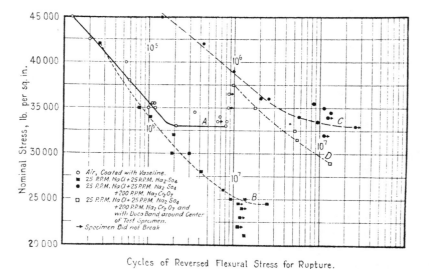

Cycles of Reversed Flexural Stress for Rupture.

Fig. 25. Effect of sodium dichromate additions to a chloride–sulphate solution on the corrosion fatigue behaviour of a mild steel. From Speller *et al.*, Ref. 107.

dichromate addition to a chloride–sulphate solution on the corrosion fatigue behaviour of mild steel[107] and indicates that when sufficient dichromate is present a significant improvement in fatigue life is possible compared with the life observed in a solution containing no inhibitor. In this case, the dichromate stimulates the formation of a chromium oxide film on the surface of the steel to enhance protection.

Recently, Agarwala and DeLuccia[108] have shown that solubilizing inorganic inhibitors into an organic phase by phase transfer catalysis enables the inhibitor to remain free of hydration shells and consequently become more effective. They demonstrated that inhibitors so prepared, containing mixtures such as dichromate, borate and nitrite, were effective in retarding crack growth rates of 4340 steel in air of 90% relative humidity by a factor of 4. However, it appears that the concentration of dissolved inhibitor required for effective resistance to corrosion fatigue is usually greater than that found to be adequate for preventing general corrosion, presumably because additional inhibitor is required to effect repair of protective films broken down by the action of the applied stress. If a sufficiently high concentration of inhibitor is not maintained then intensified action may result.[109]

REFERENCES

1. P. T. Gilbert, *Metallurgical Reviews*, **1**, 379–417 (1956).
2. L. F. Coffin, Fatigue at high temperature, ASTM STP 520, 5–34 (1973).
3. H. J. Gough and D. G. Sopwith, *J. Inst. Metals*, **49**, 93–112 (1932).
4. J. A. Roberson, *Trans, AIME*, **233**, 1799–1800 (1965).
5. K. U. Snowden, *Acta, Met.*, **12**, 295–303 (1964).
6. N. J. Wadsworth, *Internal stress and fatigue in metals*, eds Rassweiler and Grube, Elsevier, Amsterdam, pp. 382–396 (1959).
7. T. Broom and A. Nicholson, *J. Inst. Metals*, **89**, 183–190 (1961).
8. W. Engelmaier, *Trans Met. Soc. AIME*, **242**, 1713–1718 (1968).
9. H. Ishii and J. Weertman, *Scripta Met.*, **3**, 229–232 (1969).
10. M. J. Hordan and M. A. Wright, NASA Report No CR-1165, 1968.
11. R. N. Wright and A. S. Argon, *Met. Trans*, **1**, 3065–3074 (1970).
12. D. A. Meyn, *Trans, ASM*, **61**, 52–61 (1968).
13. J. C. Mabberley, RAE Tech Memo, CPM 53, 1966.
14. F. J. Bradshaw and C. Wheeler, *App. Mats. Res.*, **5**, 112–120 (1966).
15. C. M. Hudson, *Proc. of Vacuum Metallurgy Conf.*, 1972, 1424.
16. H. Kitegawa, *Corrosion fatigue, chemistry, mechanics and microstructure*, NACE 2, Eds, O. Devereux, A. J. McEvily and R. W. Staehle, pp. 521–528 (1972).
17. D. J. McAdam, Jnr, Cong Inter Lessai Materiaux, Amsterdam, pp. 305–358 (1928).
18. H. J. Gough and D. G. Sopwith, *J. Inst. Metals*, **56**, 55–89 (1935).
19. H. J. Gough and D. G. Sopwith, ibid., **72**, 415–422 (1946).
20. A. M. Binnie, *Engineering*, **128**, 190–191 (1929).
21 *Corrosion fatigue, chemistry, mechanics and microstructure*, NACE 2, eds O. Devereux, A. J. McEvily and R. W. Staehle (1972).
22. *Stress corrosion cracking and hydrogen embrittlement of iron base alloys*, NACE 5, eds R. W. Staehle, J. Hochmann, R. D. McCright and J. E. Slater (1973).

23. *Proc. Int. Conf. on the Effect of Environment on Fatigue*, 1977, Inst. Mech. Eng. London, ISBN 0 85298 376X.
24. *Mechanisms of environment sensitive cracking of materials*, 1977, The Metals Society, ISBN, 0 904357, 13 9.
25. Corrosion fatigue, 1978, *Metal Science*, **13**, ISBN, 0306–3453.
26. W. H. Bamford, *Trans, ASME, J. Eng. Mats and Tech.*, **101**, 182–190 (1979).
27. P. M. Scott, Inst, Mech. Eng. *Tolerance of flaws in pressure vessels*, London, 87–98 (1978).
28. P. M. Scott, *Mats Sci. J.*, **13**, 396–401 (1979).
29. S. T. Rolfe and J. M. Barsom, *Fracture and fatigue control in structures*, eds N. M. Newmark and W. J. Hall, Prentice Hall Inc., New Jersey (1977).
30. I. L. Mogford, *Developments in pressure vessel technology*, ed. R. W. Nichols, Vol 1, Applied Science Publ. Ltd, London (1979).
31. H. Neuber, *Theory of notch stresses*, J. W. Edwards, Ann Arbor, Mich. (1946).
32. R. E. Peterson, *Stress concentration factors*, John Wiley and Sons Inc., ISBN 0-471-68329-9 (1974).
33. M. Simnad and U. R. Evans, *Proc Roy Soc.*, **188**, 372–392 (1947).
34. D. J. Duquette and H. H. Uhlig, *Trans ASM*, **62**, 839–845 (1969).
35. C. Laird and D. J. Duquette, *Corrosion fatigue, chemistry, mechanics and microstructure*, NACE 2, pp. 88–117 (1972).
36. R. A. Olieh, PhD Thesis, 1980, University of Newcastle upon Tyne, England.
37. M. Pourbaix, *Atlas of electrochemical equilibria in aqueous solutions*, Pergamon Press, p. 174 (1966).
38. J. Congleton, R. A. Olieh and R. N. Parkins (in press).
39. R. Johnson, A. McMinn and B. Tomkins, *Mechanical behaviour of materials*, ICM 3, vol 2, pp. 371–382, eds K. J. Miller and R. F. Smith, Pergamon Press, ISBN, 0-08-024739-3 (1980).
40. A. S. Tetleman and A. J. McEvily, Jnr, *Fracture of structural materials*, John Wiley (1967).
41. J. F. Knott, *Fundamentals of fracture mechanics*, Butterworths (1973).
42. D. Broek, *Elementary engineering fracture mechanics*, Noordhoff, Leyden (1974).
43. H. M. Westergaard, *J. App Mech.*, **6**, 49–53 (1939).
44. W. Elber, Damage tolerance in aircraft structures, ASTM, STP406, p. 230 (1971).
45. J. M. Barsom, *J. Eng. Fract. Mech.*, **3**, 15–25 (1971).
46. R. P. Wei and G. W. Simmons, *Stress corrosion cracking and hydrogen embrittlement of iron base alloys*, NACE 5, 751–765 (1977).
47. P. C. Paris and F. Erdogan, *J. Basic Eng. ASME Trans.*, **85**, 528–534 (1963).
48. K. Tanaka and S. Matsuoka, *Int. J. Fract.*, **13**, 563–583 (1977).
49. K. Tanaka, *Int. J. Fract.*, **15**, 57–68 (1979).
50. J. Congleton and I. H. Craig, *Int. J. Fract.*, **16**, R207–R209 (1980).
51. B. Tomkins and W. D. Biggs, *J. Mat. Sci.*, **4**, 544–553 (1969).
52. A. J. Krasowsky and V. A. Stepanenko, *Int. J. Fract.*, **15**, 203–215 (1979).
53. P. J. E. Forsyth, *Crack propagation symposium*, Cranfield, 76–94 (1961).
54. P. J. E. Forsyth and D. A. Ryder, *Metallurgia*, **63**, 117–124 (1961).

55. P. J. E. Forsyth, C. A. Stubbington and D. Clark, *J. Inst. Metals*, **90**, 238–239 (1961–62).
56. R. J. H. Wanhill, *Corrosion*, **21**, 66–71 (1975).
57. R. O. Ritchie and J. F. Knott, *Acta Met.*, **21**, 639–648 (1973).
58. J. D. Atkinson and T. C. Lindley, The influence of environment on fatigue, Inst. Mech. Eng. Publications, **4**, 65–74 (1977).
59. J. P. Gallagher, *J. Materials*, *JMLSA*, 1971, **6**, 941–964 (1971).
60. M. O. Spiedel, *Stress corrosion cracking and hydrogen embrittlement of iron base alloys*, NACE 5, 1071–1094 (1977).
61. J. M. Barsom, *Corrosion fatigue*, NACE 2, 424–436 (1972).
62. B. K. Denton, Private communication.
63. J. D. Atkinson and T. C. Lindley, *Metal Sci. J.*, **13**, 444–448 (1979).
64. O. Vosikovsky, *Trans, ASME, J. Eng. Mats and Tech.*, **97**, 298–304 (1975).
65. P. E. Irving and C. J. Beevers, *Met. Trans.*, **5**, 391–398 (1974).
66. B. Tomkins, The influence of environment on fatigue, Inst. Mech. Eng. Conf., Publ. 4, 111–116 (1977).
67. P. Newmann, H. Fuhlrott and H. Vehoff, Fatigue mechanisms, ASTM STP 675, ed. J. T. Fong, 371–395 (1979).
68. N. Thompson, N. Wadsworth and N. Loaut, *Phil. Mag.*, **1**, 113–126 (1956).
69. M. Gell and G. R. Leverant, *Proc. 2nd Int. Conf. on Fracture*, Brighton, UK, 565–575 (1969).
70. R. B. Scarlin, *Proc. 4th Int. Conf. on Fracture*, Waterloo, Canada, **2**, 849–857 (1977).
71. R. Wearmouth and R. N. Parkins, in R. N. Parkins, F. Mazza, F. F. Royuela and J. C. Scully, *Br. Corr. J.*, **7**, 154–167 (1972).
72. T. Kunio and K. Yamed, Fatigue mechanisms, ASTM STP 675, ed. J. T. Fong, 342–370 (1979).
73. K. Miller, ibid, 361–366.
74. J. M. Krafft and W. H. Cullen, Jnr, NRL report 3505 (1977).
75. I. M. Austin and E. F. Walker, *Proc Int Conf on the Effect of Environment on Fatigue*, Inst. Mech. Eng: Conf., Publ. 4, pp 1–10 (1977).
76. R. P. Wei and J. D. Landes, *Mats Res. Standards*, **9**, 25–27 (1969).
77. F. P. Ford, *Corrosion*, **35**, 281–287 (1979).
78. D. A. Vermilyea and C. S. Tedman, *J. Electrochem. Soc.*, **117**, 437–439 (1970).
79. B. G. Ateya and H. W. Pickering, ibid, **122**, 1018–1026 (1975).
80. P. Doig and P. E. J. Flewitt, *Proc. Roy. Soc.*, **357A**, 439–452 (1977).
81. R. N. Parkins, *Metal Sci. J.*, **13**, 381–386 (1979).
82. J. M. Barsom, *Int. J. Frac. Mech.*, **7**, 163–182 (1971).
83. U. R. Evans, *Corrosion and oxidation of metals*, Edward Arnold, London, 1960.
84. B. F. Brown, The theory of stress corrosion cracking in alloys, *NATO Science Committee Research Evaluation Conference*, Brussels, 186–204 (1971).
85. M. Pourbaix, ibid, 17–63.
86. W. H. Hartt, J. S. Tenant and W. C. Hooper, Corrosion fatigue technology, ASTM STP 642, 5–18 (1978).
87. J. P. Gallagher and G. M. Sinclair, *J. Basic Eng.*, **91**, 598–602 (1969).

88. J. P. Gallagher and R. P. Wei, *Proc Conf on Corrosion Fatigue*, NACE 2, 409–423 (1972).
89. J. C. Grosskreutz, ibid, 451–452.
90. W. E. Krupp, D. W. Hoeppner and E. K. Walker, ibid, 468–483.
91. R. P. Wei, *Eng. Fract. Mech.*, **1**, 633–651 (1970).
92. P. Smith and A. T. Stuart, *Metal Sci. J.*, **13**, 429–435 (1979).
93. B. Tomkins, *Metal Sci. J.*, **13**, 387–395 (1979).
94. R. N. Parkins, ASTM STP 665, 5–25 (1979).
95. J. C. Scully and D. T. Powell, *Corrosion Sci.*, **10**, 719–733 (1970).
96. D. A. Jones, ibid, **11**, 439–451 (1971).
97. J. L. Nichols, *Materials Protection*, **2**, 46–53 (1963).
98. C. M. Hudgins, B. M. Casad, R. L. Schroeder and C. C. Patton, *J. Petroleum Technology*, **23**, 283–293 (1971).
99. B. F. Brown, *Corrosion fatigue, chemistry, mechanics and microstructure*, NACE 2, 25–29 (1972).
100. P. W. Marshall, J. W. Kochera and J. P. Traemer, Fatigue of structural steel for offshore platforms, Paper OTC 1046, *Offshore Technology Conf*, Houston (1969).
101. L. P. Pook and A. F. Greenan, *Fatigue testing and design*, Vol 2, Society of Environmental Engineers, London, 30.1–30.33 (1976).
102. W. E. Harvey, *Metals and Alloys*, **1**, 458–461 (1930).
103. I. J. Gerard and H. Sutton, *J. Inst. Metals*, **56**, 29–54 (1935).
104. D. R. Gabe, *Principles of metal surface treatment and protection*, Pergamon Press (1972).
105. F. Fancutt and J. C. Hudson, *Protective painting of structural steels*, Chapman Hall (1957).
106. W. E. Cowley, F. P. A. Robinson and J. E. Kerrich, *Br. Corr. J.*, **3**, 223–237 (1968).
107. F. N. Speller, I. B. McCorkle and P. F. Murma, *Proc ASTM*, **28**, 159–167 (1928).
108. V. S. Agarwala and J. J. DeLuccia, *Corrosion*, **36**, 208–212 (1980).
109. A. J. Gould, *Int. Conf. on Fatigue*, Inst. Mech. Eng., 341–347 (1956).
110. F. D. Bogar and T. W. Crooker, NRL Report 8153, Effects of natural seawater and electrochemical potential on fatigue crack growth in 5086 and 5456 aluminium alloys (1977).
111. T. W. Crooker, F. D. Bogar and W. R. Cares, ASTM STP 642, Corrosion fatigue technology, eds, H. L. Craig, T. W. Crooker and D. W. Holppner, pp. 189–201 (1978).
112. P. S. Pao, W. Wei and R. P. Wei, *Proceedings Symposium on Environment Sensitive Fracture of Eng Mats*, TSM-AIME (1979).
113. M. O. Spiedel, M. J. Blackburn, T. R. Beck and J. A. Feeney, *Corrosion fatigue, chemistry, mechanics and microstructure*, eds O. Devereux, A. J. McEvily and R. W. Staehle (1972).
114. P. M. Scott and D. R. V. Silvester, Harwell Interim Technical Reports, UKOSRP 3/02; 3/03, 1975 and 1977.
115. W. H. Cullen, V. Provenzano, K. J. Torronen, H. E. Watson and F. J. Loss, NUREG/CRO969, NRL Memorandum Report 4063 (1979).
116. W. H. Bamford and D. M. Moon, *Corrosion*, **36**, 289–298 (1980).

117. F. J. Loss, NUREG/CR-1268, NRL Memorandum Report 4174 (1980).
118. J. T. Ryder and J. P. Gallagher, *J. Test. and Mat. Eval.*, **2**, 180–189 (1974).
119. A. M. Sullivan and T. W. Crooker, NRL Report 7936 (1975).
120. S. Aiyama, T. Shoji, H. Takahashi and M. Suzuki, *Corrosion*, **10**, 325–330 (1978).
121. B. Mukherjee, ASTM STP 642, Corrosion fatigue technology, pp. 264–285 eds. H. L. Craig, T. W. Crooker and D. W. Holppner (1978).
122. T. W. Crooker and E. A. Lange, 1969, *J. Basic Eng. Trans ASME*, **91**, pp. 570–580.
123. C. Amzallag, P. Rabbe and A. Destret, ASTM STP 642, Corrosion fatigue technology, pp. 117–132, eds H. L. Craig, T. W. Crooker and D. W. Holppner (1978).
124. J. A. Moskovitz and R. M. Pelloux, ibid, pp. 133–154.
125. R. P. Jewett, R. J. Walter and W. T. Chandler, ibid, pp. 243–263.

Chapter 6

STRESS CORROSION CRACKING

F. P. FORD

General Electric Co., Research and Development Center,
New York, USA

1. INTRODUCTION

|Stress corrosion cracking is a phenomenon associated with a combination of nominally static tensile stress, environment and, in some systems, metallurgical condition, which leads to component failure due to the initiation and propagation of a high aspect ratio crack.|Cracking may be either intergranular or transgranular and may progress at propagation rates in a wide range 10^{-9} mm s^{-1} to 10^{-1} mm s^{-1}.

Similar criteria are required to a greater or lesser extent, for corrosion fatigue, hydrogen embrittlement or liquid metal embrittlement although in the last two situations, failure can be associated with lack of tensile ductility rather than relatively slow growing cracks; these overall similarities have led to attempts to formulate common theories of environmentally controlled cracking.[1-3]

The materials involved in these phenomena may be metallic or non-metallic and the aggressive environments may be aqueous (organic or inorganic), gaseous or metallic. In this chapter attention is focused specifically on stress corrosion cracking in ductile alloy/aqueous environment systems, how this phenomenon relates to hydrogen embrittlement and corrosion fatigue and, in particular, how recent advances in the understanding of the mechanisms of stress corrosion can help in the formulation of remedial actions and design criteria.

The advances in mechanistic understanding of stress corrosion have occurred in three distinct phases.[4] The first two were an 'identification' phase in the 1940s and 1950s in which the problem was identified and categorized in terms of specific 'alloy/aggressive anion' couples (Table 1), and a 'mechanistic' phase in the 1960s and early 1970s in which various mechanisms were proposed to explain the crack morphology, the dependency of cracking on various electrochemical and metallurgical parameters and the wide range of crack propagation rates observed. It was during this latter phase, following the improvement in sensitivity of crack-following techniques required to evaluate alloys for components with, say, 40 year design lives, and the concentrated investigation of many other alloy/environment systems, that it was realized that the list of damaging species in Table 1 was by no means complete, that 'pure' metals as well as alloys could crack, and that cracking was possible for many alloys in pure water.

TABLE 1
STRESS CORROSION SYSTEMS RECOGNIZED *c*. 1960–1965

Mild steel	OH^-, NO_3^-, NH_4^+, CO/CO_2, CO_3^{2-}/HCO_3^-
Low alloy ferritic steel	
Austenitic stainless steel	Cl^-, OH^-, $H_2O(O_2)$
Aluminum alloys	Cl^-
Copper alloys	NH_4^+
High strength steels	H_2O, $H_{2(g)}$, $H_2S_{(g)}$
Titanium alloys	Cl^-, NO_3^-, fused salts, CH_3OH

This alarming situation (from a technological viewpoint) led to the present third phase involving the application of mechanistic knowledge to practical needs; specifically in improving predictive capabilities through knowledge of rate-determining steps in the cracking mechanism, in giving guidelines to the development of material compositions or environment control which improved cracking resistance, in developing testing techniques which were relevant to practice and could incorporate their results into a design criterion, and in identifying the actual environmental and material conditions present in components which could potentially suffer cracking damage. (This latter item is not covered in the following discussion, although it is important to realize that in many failure analyses or risk assessments, the actual alloy/environment system in an occluded region of a component is not always adequately or easily defined.)

2. RECENT ADVANCES IN MECHANISTIC KNOWLEDGE AND PREDICTIVE CAPABILITIES

2.1 Historical Background

The advance in mechanistic understanding of environmentally controlled phenomena during the 1960s and early 1970s has been documented in the proceedings of a series of conferences.[5-15]

Three broad categories of stress corrosion mechanisms emerged:[16] pre-existing active path mechanisms, strain-assisted active path mechanisms and adsorption-related phenomena. The essential features of these mechanisms are reviewed briefly below, since they form the basis the recent advances in fundamental understanding of the practical problem.

2.1.1 Pre-existing Active Path Mechanisms

These encompass the early mechanistic ideas and were applied[17-19] primarily to intergranular cracking of ductile alloys in aqueous environments, relating the cracking susceptibility to the chemical activity of the grain boundary. The consequent localized intergranular attack (IGA) was due to either the inherently higher activity of the disordered grain boundary structure or, more generally, to the presence of chemically active grain boundary impurities, precipitates (e.g. β phase in Al/Mg alloys) or the solute denuded zone adjacent to the precipitate (e.g. Al/Cu alloys, 18% Cr/Ni stainless steels). The pre-existing active path mechanism had support in that cracking susceptibility could be altered either by metallurgically controlling the grain boundary precipitate volume (i.e. anode/cathode area ratios) by, for instance, alloying (e.g. Zn to Al/Mg alloys) or thermomechanical treatment, or by controlling the environmental conditions to make the grain boundary region less active with respect to the adjacent matrix (e.g. pH control in aluminum alloys).

It subsequently became apparent[16] that the cracking susceptibility within the range of systems covered by the active path mechanism was controlled at one limit by predominantly localized corrosion mechanisms, where stress was unnecessary for complete component failure (e.g. mild steel/nitrate and sensitized stainless steel/oxalic acid systems at more positive potentials) to the other limit where stress had a specific role in deciding whether an IGA 'notch' progressed to a propagating crack. This latter limit was bounded by the strain-assisted active path mechanisms.

2.1.2 Strain-Assisted Active Path Mechanisms

Various crack advancement theories have been proposed relating crack propagation to dissolution at the crack tip and the stress/strain condition in that region. For instance the tensile ligament theory[20] for higher strength alloys proposed that catastrophic crack advance in absence of an environment was limited by the rupture of tensile ligaments across the crack tip opening, and that the introduction of an aqueous environment reduced the load bearing area of the ligaments by dissolution, thereby lowering the stress intensity required for subcritical crack advance. Alternatively Hoar and co-workers[21-27] had proposed for ductile alloys that dissolution at a film-free crack tip could be enhanced by the strain concentration because of an increase in active site density, e.g. kink sites, or (less likely)[28] a decrease in the activation enthalpy for dissolution. This original idea was supported somewhat by TEM evidence that mobile dislocations (expected at the crack tip) could be preferentially dissolved[29-31] due to either the inherent chemical activity of the dislocation core or to solute segregation there, and these observations gave a possible understanding of the crystallographic nature of transgranular stress corrosion and the correlation of cracking susceptibility in aqueous solutions with stacking fault energy (and consequent dislocation morphology) in several systems.[29,32-39]

The above strain-assisted active path theories, which relied on either the chemical activity of a strained region at the crack tip through plasticity arguments or on the effect of dissolution on the mechanical integrity of the crack tip, were superseded by theories[40-48] that relied on the rupture of a protective film at the crack tip. Thus it was realized that for a high aspect ratio crack to exist the sides of the crack must be protected, even through the crack tip was in a chemically active condition due, for instance, to the mechanical rupture of the protective film followed by metal penetration by bare-surface dissolution and/or film growth, hence the slip dissolution[42] and brittle film[49-53] theories were propounded. Different types of protective film were proposed varying from oxide, through mixed oxides and salts[54] to noble metals[55] left on the surface after selective dissolution of a more active component in an alloy. These theories were supported by the observation that high stress corrosion susceptibility was encountered under potential/pH conditions where a protective film was thermodynamically stable but, if ruptured, bare-surface dissolution was thermodynamically possible. Further, the observed crack propagation rates for many ductile alloy/aqueous environment systems were in direct proportion to the experimentally determined dissolution rates under the mechanical and chemical conditions expected at the crack tip (Fig. 1).[56]

2.1.3 Adsorption-Related Phenomena

Alternative mechanisms for subcritical crack propagation rely primarily on the decrease in mechanical integrity at the crack tip due to adsorption of specific species from the environment followed by possible absorption into the underlying matrix. Various mechanisms for this mechanical property degradation have been proposed.[8-11,57,58]

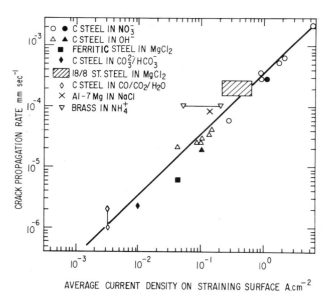

FIG. 1. Relationship between the average crack propagation rate and the oxidation (i.e. dissolution and oxide growth) kinetics on a straining surface, for several ductile alloy aqueous environment systems.[56]

Surface energy reduction models[59-63] rely on surface chemisorption of an environmental species on the crack tip, which reduces the surface energy term, γ_s, in the equilibrium Griffith relationship[64] and thereby reduces the local fracture stress of the metal lattice. Such an argument has been used for hydrogen embrittlement, liquid metal embrittlement,[62,65-67] and for the effect of specific anions in aqueous solutions[68] where the adsorbed atom coverage will depend on electrode potential.

Decohesion models[69-72] have been proposed specifically for hydrogen embrittlement and temper embrittlement, where subsurface atom–atom rupture is facilitated by local concentrations of hydrogen or temper-embrittling elements which modify the electron d-band structure or expand the lattice. Such effects can lead to a loss in tensile ductility following hydrogen charging (either thermally or electrochemically) or from heat treatments which allow segregation of, for instance, temper-embrittling elements to the grain boundary. Indeed synergistic effects have been suggested between hydrogen embrittlement and temper embrittlement in high-strength steels.[73]

It has been argued[70] that the difference in mechanism of atom–atom rupture between the surface-energy and decohesion models is minor; however the main conceptual difference is that, in the former case, rupture occurs at the crack tip surface whereas, in the latter case, subcritical crack *propagation* relies on the diffusion of the 'aggressive' atom to some point beneath the surface where rupture occurs. Such subcritical crack propagation, therefore, is specific to hydrogen (or environments where hydrogen atoms can be produced at the crack-tip surface by H^+ reduction, H_2 dissociation, etc.) because of the high mobility of the atom. The propagation process is envisaged as a discontinuous cycle of: production of adsorbed hydrogen atoms at the crack tip; surface diffusion; absorption and matrix diffusion to a region in front of the crack tip where localized mechanical fracture occurs when the hydrogen content reaches a critical value, C^*, over a critical volume, d^*.[74,75] The latter criterion, d^*, may be related to metallurgical features, e.g. grain size,[76,77] etc., but the critical hydrogen concentration, C^*, required for local fracture remains ill-defined and indeed, alternative criteria to the decohesion model for localized mechanical fracture have been proposed. For instance, that pressurized pockets of hydrogen gas form at internal discontinuities due to hydrogen atom recombination,[78-82] that brittle phases (e.g. hydrides,[83-85] hydrogen- or strain-induced martensite, etc.[86-90]) can be formed, or else that local fracture criteria are altered due to a modification of the local plasticity[91-94] by the hydrogen concentration.

Recently, it has been proposed that hydrogen embrittlement models may be operating in lower strength alloys having many active slip systems. The experimental evidence for this is wide ranging and has been reviewed by Thompson and Bernstein,[95] who indicate that hydrogen diffusion can be markedly accelerated[96-98] by dislocation transport thereby leading to hydrogen supersaturation, under suitable dynamic equilibrium conditions, at various sites, causing local fracture due to brittle phase formation, restricted plasticity, etc., and explaining the different fractographic features observed, e.g. dimple, cleavage, grain-boundary fracture, etc. (Fig. 2).

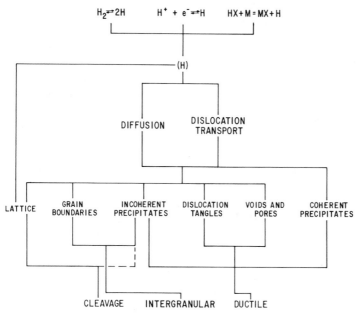

FIG. 2. Possible interaction paths for the transport of hydrogen atoms produced at the crack tip to regions in the matrix where atom–atom rupture occurs.[95]

However it is also apparent that the effect of many of the metallurgical variables on crack propagation in ductile alloy/aqueous environment systems may be explained by either the modified non-equilibrium hydrogen-transport model or by the strain-assisted active path mechanisms, e.g. coplanar dislocation morphologies would be expected to accelerate hydrogen transport as well as facilitating oxide rupture in a slip-dissolution or brittle-film mechanism. As pointed out by Parkins[16] the controlling factors in the active path, strain-assisted active path and adsorption-related propagation models form a spectrum of interrelated effects, and although under a particular testing condition one mechanism may be valid, an alteration of system variables (e.g. pH, potential, strength level, etc.) may well lead to a different mechanism being appropriate, or, alternatively, situations may exist where two mechanisms may be operating simultaneously. This latter possibility has led to much confusion in attempts to ascribe a singular mechanism to a given cracking system, the confusion being compounded by the fact that variables giving enhanced dissolution must also increase the hydrogen ion reduction rates in adjacent regions in order to maintain electroneutrality, and that (as will be discussed in detail below) the prime crack advancement alternatives may have the same rate-determining steps.

Recent advances in the environmental cracking art have concentrated on refining the details of the various mechanisms, relating the mechanisms of cracking in one class of material to that in another,[13] e.g. metals v. non-metals, understanding the surface physics of metal/environment interfaces,[12] and, possibly more important, converting the mechanistic knowledge to the solution of practical problems by both evolving prediction capabilities and evaluating remedial actions, design criteria and associated testing techniques.

2.2 Recent Advances in Understanding of Crack Propagation Mechanisms

For many systems[7,8,99] the subcritical crack propagation rate, V, has a characteristic relationship with the applied stress intensity, K (Fig. 3), where crack propagation does not occur at a discernible rate below K_{ISCC} (or K_{TH}), but rises rapidly with stress intensity in a Stage I region; at higher stress intensities, in the Stage II region, the (dV/dK) value decreases and may become zero. Very similar crack propagation rate/applied strain rate relationships[246,247] have been observed, with negligible crack growth occurring below a critical 'crack tip' strain rate. Mechanistic investigations have centered on defining the fundamental parameters of importance in these relationships in various environments.

2.2.1 Thermodynamic Considerations

Thermodynamic requirements for dissolution-based models. For a high aspect ratio crack to advance in aqueous environments by a dissolution-based mechanism it is necessary for not only dissolution to be thermodynamically possible but that a protective film (either oxide, mixed oxide, salt, etc.) also be thermodynamically stable to prevent the degradation of the crack into a pit or blunt notch[27,43,47] due to undue corrosion of the crack sides. Such requirements for an 'electrochemical knife'[43] limit the environmental conditions in which severe susceptibility

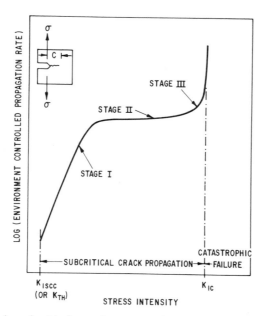

FIG. 3. Typical, subcritical crack propagation rate versus stress intensity relationship. Stress intensity K, is defined as:

$$K = A\sigma \sqrt{\frac{\pi C}{B}}$$

where σ = total tensile stress
C = crack length
A, B = geometrical constants

is possible, and provide a predictive capability for identifying the potential/pH combinations where danger situations may exist in practice. For instance, cracking of mild steel in aqueous environments (Fig. 4) by strain-assisted active path mechanisms is confined to potential/pH regions where a soluble species (Fe^{2+}, $HFeO_2^-$) can form when a protective magnetite, mixed oxide or salt film in OH^-, NO_3^-, CO_3^{2-} or PO_4^{3-} containing solutions is ruptured. Such an approach to the prediction of environmental conditions in which severe cracking is possible has also been applied to non-ferrous systems, e.g. brasses in ammoniacal solutions.[100]

An energy balance argument[101,102] has been used to define the stress intensity required to initiate crack propagation by dissolution-controlled

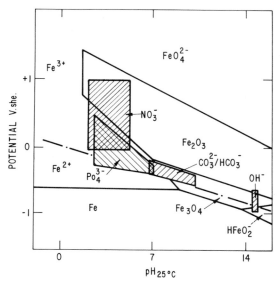

FIG. 4. Relationship between pH/potential conditions for severe cracking susceptibility of mild steel in various environments and the stability regions for solid and dissolved species on the potential–pH diagram. Note that severe susceptibility is encountered where a protective film (phosphate, mixed carbonate, magnetite, etc.) is thermodynamically stable, but if ruptured, a soluble species $(Fe^{2+}, HFeO_2)$ is metastable.

mechanisms, on the basis that an equilibrium thermodynamic approach defines the condition when the crack propagation rate is zero. In this treatment the Griffith[64]–Orowan[103]–Irwin[104] criterion for mechanical fracture has been expanded by summating the elastic strain-energy release rate and the chemical-energy change and equating them with the increase in system energy due to changes in total surface energy and plastic energy accompanying an incremental increase in crack length. The decrease in system energy associated with chemical processes is equated with dissolution at a continuously bared crack tip. The K_{ISCC} values calculated by this method are markedly lower than those experimentally observed,[102] the discrepancy arising out of the assumption in the calculations that the crack tip is always in a bared condition. Although closer agreements between observed and calculated K_{ISCC} values[105] can be obtained by recognizing that a primary requirement for cracking is the *rupture* of the surface film, the practical importance of the energy balance approach is the indication that any operation which forces the crack tip towards a continuously bared condition, e.g. dynamic or cyclic loading, should give rise to a monotonic lowering of the stress intensity above which crack propagation is observed. This aspect of the spectrum between 'stress corrosion' and 'corrosion fatigue' for systems where rupture of oxide at the crack tip is a governing feature is discussed in more detail in Section 2.2.2.

Thermodynamic requirements for hydrogen-embrittlement-related models. Similar thermodynamic reasoning has been applied to the definition of the environmental and stress conditions for various hydrogen-embrittlement models. For instance the environmental criterion in aqueous conditions that hydrogen ions be reduced to hydrogen adatoms at the crack tip is definable on the Pourbaix equilibrium diagram.

Theoretical considerations of K_{TH} have included the application of deformation kinetics theory,[106,107] the consideration of variations in surface energy terms[108] and of the stress conditions governing the achievement of a critical hydrogen concentration beneath the film-free crack surface.[109] This latter approach has used the thermodynamic relationship between the equilibrium hydrogen solubility and the dilational stress in front of the crack tip.[110,111] By equating the onset of subcritical crack propagation with the achievement of a critical hydrogen concentration, redefining the stress in terms of stress intensity and considering the stress distribution in front of a crack as a result of small-scale yielding, Gerberich and co-workers[76,109,112] were able to obtain satisfactory predictive relationships between K_{TH} (or K_{ISCC}) and such parameters as yield stress, stressing mode (plane strain or plane stress conditions, grain size, etc.). Other workers[74,75] have extended this approach to the requirement that the critical hydrogen concentration be exceeded over a specific volume in front of the crack, thereby introducing predictive capabilities for the effect of blunt notches, initiation times, etc. It should be noted however that these models have been applied to higher strength alloys where non-equilibrium effects such as dislocation-aided hydrogen transport are minimal; corresponding predictive capabilities for K_{TH} in hydrogen embrittlement of more ductile alloys have not yet been achieved.

Thermodynamic determination of the validity of a given cracking mechanism. The viability of a given cracking mechanism may be assessed on the basis of the satisfaction of the thermodynamic requirements for that model under the appropriate alloy/environment conditions. For instance,

Parkins *et al.*[113] were able to predict correctly the regions of cracking in the mild steel/phosphate system due to either a strain-assisted active path or a hydrogen-embrittlement mechanism, an ability of practical importance with respect to the limitations of remedial actions based on either environmental control or cathodic protection. Similarly on the basis of the predictability of K_{TH}/yield strength relationships[109] for hydrogen embrittlement (Fig. 5(a)) it could be argued that cracking of lower strength steels in aqueous environments *cannot* be due to such a mech-

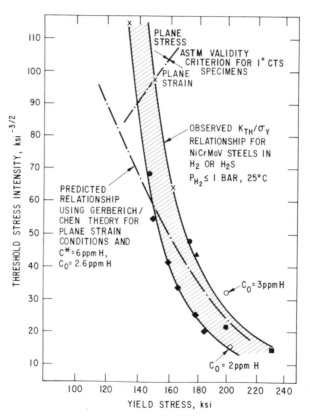

FIG. 5(a). Comparison between observed and predicted K_{TH}/yield stress relationships for NiCrMoV steels in hydrogen or hydrogen sulphide. Note the good correlation under plane strain conditions. Predicted values based on theory of Gerberich and Chen[109] using values of $C^* = 6\,ppm\,H$ and the hydrogen content in the unstressed condition, C_0, of 2.6 ppm H.

anism since the observed K_{ISCC} values[114,115] are markedly lower than the minimum predicted value (Fig. 5(b)) and that a strain-assisted active path mechanism such as slip dissolution with its low predicted K_{ISCC} values would be more appropriate. However this conclusion does not take into account non-equilibrium hydrogen-transport mechanisms which are possible in ductile alloys, and that, even in bulk neutral solutions where it could be argued that hydrogen ion reduction is not thermodynamically possible at more positive potentials, e.g. aerated

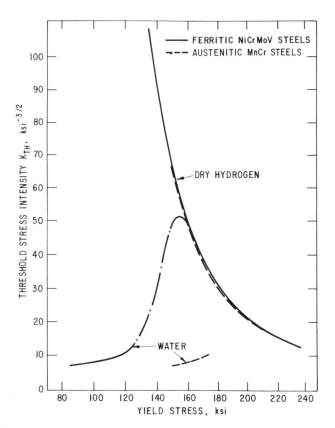

FIG. 5(b). Comparison between the K_{TH}/yield stress relationships for NiCrMoV ferritic steels and MnCr austenitic steels in dry hydrogen and in water. Note the markedly lower K_{ISCC} (or K_{TH}) values in water or at lower yield strengths, but the similar threshold values in the two environments at higher yield strengths.

solutions, it has been demonstrated[116-119] that the potential/pH conditions at the crack tip *are* thermodynamically conducive to a hydrogen-embrittlement mechanism.

In reviewing recent advances in cracking predictability and mechanism differentiation based on thermodynamic considerations alone it is apparent that although in certain circumstances it is possible to identify the likely cracking mechanism by examination of the cracking susceptibility in relation to the potential–pH diagram or the equilibrium stress requirements (Fig. 5(a)), in many practical cases involving ductile alloys in aqueous environments both strain-assisted active-path and hydrogen-embrittlement models are possible under particular environmental or stress conditions. To a large extent this ambiguity stems from lack of quantitative knowledge of the non-equilibrium considerations associated with hydrogen-transport mechanisms and the uncertain definition of the potential and environment at the crack tip. This latter point is referred to again in Section 2.3.

2.2.2 Kinetic Considerations

Arguable attempts[101,120-124] have been made to explain subcritical crack growth rates in stress corrosion and corrosion fatigue by inserting extra chemical energy terms in the *equilibrium* Griffith–Orowan energy balance criterion for the stability of cracks. However more complete explanations of the V/K (Fig. 3) relationship for ductile alloy/aqueous environment systems have equated the stress-dependent Stage I cracking with an increasing plastic strain rate at the crack tip and associated oxide rupture rate and/or dislocation-aided hydrogen transport rate. The subsequent stress-independent Stage II cracking has been equated with a change in rate-determining step, e.g. to liquid diffusion processes associated with either removing solvated cations from the crack tip or introducing solvating water molecules to the crack tip, or to a change in the oxide rupture rate/stress intensity relationship due to elastic/plastic stress/strain considerations in front of the crack tip.

Advances in the mechanistic understanding of the kinetics of cracking have been associated with quantitatively defining the limits and rate-determining steps in Stages I and II crack propagation.

Determination of maximum crack propagation rates, V_{max}. The maximum crack propagation rate by the slip-dissolution mechanism may be fundamentally equated with the dissolution rate at the crack tip when it is *maintained* in a bare condition.

Such a prediction of V_{max} has been performed for several systems using values of the dissolution rate obtained after rapidly fracturing or removing the surface oxide by techniques such as straining,[125-128] fracturing the whole specimen,[129] scratching,[130-132] abrading[133,134] or cathodic cleaning. The calculated maximum propagation rate follows an activation-controlled (Tafel) relationship at low dissolution overpotentials whereas at higher (and more practically relevant) overpotentials, the calculated maximum propagation rate is controlled by liquid diffusion considerations preparatory to the precipitation of oxides or salts whose solubility product has been exceeded.

Analogous maximum propagation rate predictions for non-equilibrium hydrogen transport models of hydrogen embrittlement are hampered by lack of quantitative knowledge of the hydrogen adatom coverage at the crack tip, the subsequent dynamic equilibrium between dislocation-aided hydrogen diffusion to a trapping site and transport away from the supersaturated site, and specific details of the final atom–atom rupture event. Because of these limitations, calculations[131] of the maximum crack propagation rate by the hydrogen-embrittlement model have had to rely on the *equilibrium* relationships between the hydrogen activity and both the electrode potential at the crack tip[135] and the hydrostatic stress immediately below the crack tip surface.[110,111] Such an assumption of the applicability of equilibrium thermodynamics to a dynamic crack propagation situation is obviously questionable.

However, the maximum calculated propagation rates[131] for either basic advancement mechanism are rarely observed experimentally[127,136,137] under static loading conditions (Fig. 6), since under these stress conditions a bare-surface condition is *not* continuously maintained at the crack tip. Thus, it is apparent that at potentials of practical interest, e.g. corrosion potentials in the range -1000 mV to -800 mV sce in Fig. 6, *both* hydrogen-embrittlement and slip-dissolution mechanisms are kinetically and thermodynamically viable and may, as has been proposed for austenitic stainless steel,[138,139] aluminum alloys[140-142] or titanium alloys,[143] be operating simultaneously at the crack tip.

The question arises as to whether this lack of quantitative discrimination ability based on an examination of the maximum possible crack propagation rates is of immediate practical importance, especially when the observed propagation rates are considerably lower than these maximum theoretical values.

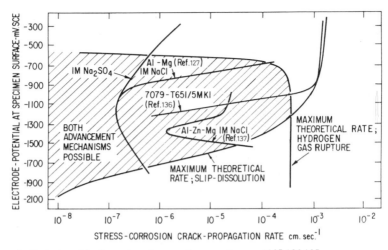

FIG. 6. Relationship between the maximum observed[127,136,137] stress corrosion crack propagation rates for various aluminum alloy/aqueous environment systems and the maximum theoretical propagation rates for the slip-dissolution and hydrogen embrittlement (gas-rupture) theories.[131] Note the fact that in the potential range -1000 to $-800\,mV$ sce (covering the corrosion potential for aerated or slightly deaerated solutions) both advancement mechanisms are kinetically viable.

Determination of rate-determining step for propagation rates less than the maximum theoretical rate. Examination of the possible rate-determining steps for hydrogen-embrittlement or slip-dissolution crack-advancement models in ductile alloy/aqueous environment systems indicates that there are many common interrelated steps (Fig. 7) in these electrochemically controlled mechanisms. For instance, for a given condition at the crack tip of potential, pH and anion content, both mechanisms will be dependent on liquid diffusion rates, passivation rates and oxide rupture rates, since these factors will affect the charge passed per unit time in the slip-dissolution model and the hydrogen adatom coverage and subsequent hydrogen permeation rate in the hydrogen-embrittlement model. Advances in understanding crack propagation processes and the numerous anomalies associated with them have arisen from knowledge of the interrelationships shown in Fig. 7.

Passivation rate. The importance of passivation kinetics on crack propagation has been increasingly recognized.[42–45,134,144–151] Slow

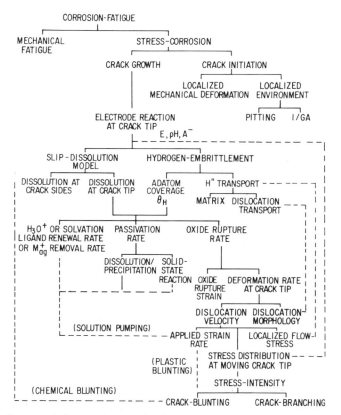

FIG. 7. Suggested interaction paths between the modes and mechanisms of environmentally controlled fracture in ductile alloy/aqueous environment systems and the various rate-determining steps in crack propagation. Solid lines denote primary interactions, dashed lines denote secondary interactions.

passivation rates at the crack tip will promote crack blunting due to excessive dissolution on the crack sides, whereas very fast rates will minimize the amount of crack tip penetration per oxide-rupture event; maximum susceptibility, with high aspect ratio cracks, will occur at intermediate passivation rates. The effects of potential, anion (or cation) content and alloying addition on cracking susceptibility may be quantitatively understood by this simple concept, regardless of whether the advancement mechanism is slip dissolution or hydrogen embrittlement. For instance, cracking susceptibility in poorly passivating systems (e.g.

austenitic stainless steel in caustic at high temperature, mild steel in caustic, phosphate, etc.) will be increased by actions which promote passivation. Thus in these systems cracking susceptibility will be greatest in potential ranges adjacent to active/passive transitions on a polarization curve (Fig. 8(a)), whereas systems which exhibit strongly passivating behavior (aluminum alloys, austenitic stainless steels in neutral solutions) will crack most severely under potential conditions where incipient passivity breakdown occurs due to the presence of aggressive anions (e.g. chloride) (Fig. 8(b)).

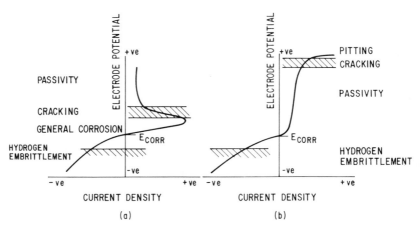

FIG. 8. Schematic electrode potential/current density relationships for (a) 'poorly' passivating and (b) 'strongly' passivating systems, indicating in the two cases where severe cracking susceptibility in ductile alloy/aqueous environment systems is commonly encountered.

Traditionally, this approach for identifying the potential range of maximum cracking susceptibility has been applied to the slip-dissolution mechanism at potentials where the thermodynamic criteria for this model are met, while the cracking at more negative potentials (Fig. 8) has been associated with hydrogen embrittlement when proton or water reduction is thermodynamically possible and the adatom coverage is high.

However it is possible that, if localized acidity at the crack tip exists due to hydrolysis of the metal cations, proton reduction leading to hydrogen adsorption and absorption can occur there even though the external surface potential is positive to the reversible potential for the

H^+/H reaction associated with the bulk pH^{118} and that the hydrogen adatom coverage at the crack tip and subsequent absorption rate will be dependent on the degree of passivation.[152-155]

This realization of the common controlling process of passivation presents a caution to the use of cracking susceptibility/electrode potential relationships as sole evidence for either the slip-dissolution or hydrogen-embrittlement mechanisms, since it can be demonstrated[131] that these relationships are due to the potential dependence of the rate-determining step, not the crack advancement reaction itself.

It is apparent, therefore, that cracking susceptibility can be minimized by control of the rate-determining reaction and, although there is an academic need to identify the crack advancement mechanism, e.g. slip dissolution or hydrogen embrittlement, this need may be of minor importance practically, especially in cases where the rate-determining reaction is common to both advancement mechanisms. Thus any change in operating condition which alters the corrosion potential so that it enters the range where the passivation rate is such as to maintain a high aspect ratio crack but allow appreciable crack tip penetration or hydrogen absorption during each oxide rupture event, will increase the cracking susceptibility.[151,156-159] For example, alteration of the corrosion potential in the positive direction due to, for instance, the presence of oxidizing species (O_2, H_2O_2, dichromates, Fe^{3+}, etc.) is deleterious for the strongly passivating systems in chloride solutions (Fig. 8(b)), e.g. aeration and its deleterious effect on chloride cracking of aluminum alloys[160] and austenitic stainless steels;[161,162] similar effects are noted in poorly passivating systems, e.g. the deleterious effect of PbO additions to the mild steel/NaOH systems[163] (Fig. 8(a)). Alloying additions may also have a similar effect due to the corrosion potential/ cracking potential range congruency. For instance, the deleterious effect[151,164] of Pt alloying to 304 austenitic stainless steel in acidic chloride solutions through its acceleration of the recombination step in the hydrogen reduction process and the consequent movement of the corrosion potential in the positive direction (Fig. 8(b)). Conversely environmental additions which decrease the hydrogen ion reduction rate, and thereby move the corrosion potential in the negative direction, may decrease the danger of cracking in passivating systems, e.g. Pb^{2+}[165] or Sn^{2+}[166] additions to the 304 stainless steel/$MgCl_2$ system.

Alternatively, alloy or environmental variables at a constant potential may improve the cracking resistance through a direct effect of increasing the passivation rate, e.g. silicates in mild steel/NaOH,[163,167] nitrates in aluminum alloy/chloride,[168] mild steel in nitrite or nitrate solutions,[134] aromatic inhibitors in stainless steel/chloride[169] systems, etc.

The fundamental importance of passivation on the crack propagation process in ductile alloy/aqueous environment systems also indicates an analytical method of determining the potential ranges where cracking susceptibility may be severe; for instance, by the direct measurement of passivation rate,[170] by comparing the 'bare surface' and 'fully passivated' dissolution rates[125,126,171] to determine whether a high aspect ratio crack is possible, or by performing potentiodynamic scans at various rates.[193] Such rapid prediction capabilities are of use in preliminary failure analyses or risk assessments, but this usefulness should be tempered by the realization that these techniques will indicate only the possibility of severe susceptibility. For the prediction of systems of marginal suscepti- bility (but significant over long design lives) full consideration is neces- sary of all the possible rate-determining steps, e.g. passivation rates *and* diffusion and oxide rupture rate effects.

Diffusion rates (Fig. 7). The main experimental evidence for liquid diffusion control in ductile alloy/aqueous environment systems during Stage II propagation is the observation in a limited number of systems of low activation enthalpies[136,172] approximating to those independently observed for liquid diffusion, and the dependency of Stage II crack growth in aluminum[136] and titanium[54] alloys on liquid viscosity.

Various diffusional analyses[43,173-176] have been performed to relate the diffusion rate of various species in the liquid to the propagation rate and the reaction order. The nature of the species of importance is, however, in question; for instance, the diffusion of solvating water molecules or hydrogen ions to the crack tip[131] may be of importance although consideration of the diffusion of solvated metal cations away from the crack tip is noteworthy even through these species are on the 'product' side of a kinetic equation. This latter consideration arises out of the possibility that the solubility product of an oxide or salt may be exceeded if the solvated metal cation removal rate is not high enough, thereby stifling further accelerated dissolution or hydrogen ion reduction at the crack tip. Alternatively it has been proposed for aluminum[136] and titanium[43,54] alloys in chloride solutions that it is the diffusion of chloride anions to the crack tip which is rate controlling through the requirement that the salt be precipitated *adjacent* to the crack tip, thereby maintaining a high-aspect ratio crack; conversely, it could be argued that chloride diffusion is controlling the crack propagation through its adsorption rate into the oxide at the crack tip and subsequent effect on the passivation rate.

Thus, although there is little doubt that liquid diffusion is playing a dominant role in controlling the stress-independent crack propagation rate, the precise quantitative interpretation and understanding of this role remains unclear for the majority of ductile alloy/aqueous environment systems.

Oxide rupture rate (Fig. 7). Although the use of fracture mechanics parameters such as stress intensity, crack-opening displacement, etc., are useful in the design sense, it is seen from Fig. 7 that they are probably of secondary importance when considering the fundamental controlling parameters in cracking of ductile alloy/aqueous environment systems, and that their importance lies in their effect on the oxide (or film) rupture event, the subsequent oxide (or film) rupture rate and the ease of solution flow down the crack length.

The preliminary requirement for crack advance by either the 'slip-dissolution', 'brittle-film' or hydrogen-embrittlement mechanism is the rupture of the protective film at the crack tip.[105] Thus, in most systems, a primary requirement for stress corrosion cracking with respect to the mechanical properties of the surface film is that it be relatively brittle under the complex stress conditions at the crack tip, and if for some reason it becomes ductile, cracking will cease.[48,105] As a side comment, it should be mentioned that it has been culculated[105] that crack advance through *thinning* of a ductile film by a homogeneous crack-tip strain and the subsequent growth of the film into the metal substrate to its equilibrium thickness cannot account for observed crack growth rates unless very high strain rates are operating at the crack tip, or a thick oxide can rapidly form; however, such an analysis, which assumes a uniform crack tip strain, rather than the inhomogeneous strain distribution associated with slip step emergence, must be overly simplistic.

The oxide rupture prerequisite for crack advance may be associated with the interrelationship between cracking susceptibility and slip morphology, since coarse slip will be more likely to rupture a brittle film of given thickness and expose bare metal, than fine slip.[156] As mentioned briefly previously such a relationship has been observed in a variety of alloy environment systems (e.g. austenitic stainless steel in aqueous[177,178] and hydrogen[93,94,98] environments), the different dislocation morphologies being related to changes in stacking-fault energy,[29,178,179] short-range order,[140,180,181] precipitate/matrix coherency and precipitate distribution in aluminum and nickel superalloys[95,182-184] caused by overaging or thermomechanical treatment. It should be emphasized however that although dislocation planarity may facilitate the initiation of cracking, anomalies to the dislocation morphology/cracking susceptibility relationship exist, since rupture of the crack-tip film is not the *sole* requirement for subsequent crack propagation. Thus the achievement of dislocation planarity is not a *sufficient* reason for hydrogen embrittlement or stress corrosion susceptibility[37,39,178,185,186] in ductile alloys.

Vermilyea[105] has made quantitative predictions of K_{ISCC} for ductile alloy/aqueous environment systems based on the primary need to rupture the film at the crack tip, through examination of the stress/strain conditions at the crack tip and how the strain in the oxide alters as the crack advances by corrosion. Such an approach yields calculated K_{ISCC} values close to the experimental data and predicts the effects of variations in plane strain/plane stress conditions, degree of strain hardening, and corrosion rate, etc.

However, Scully[187] has pointed out that such an initiation criterion based solely on the attainment of the fracture strain, ε_f, of the surface oxide must be incomplete, since cracking is not observed below a critical value of applied strain *rate* in several systems[188-195] (where eventually ε_f *must* be exceeded). It is assumed therefore that the oxide can be ruptured on the initial application of load, but the criterion for subsequent crack growth depends on the *continued* rupture of the oxide which reforms at the crack tip. This latter criterion will depend, for instance, on the activation of dislocation sources by the stress/strain field associated with the crack tip as it advances following the initial oxide rupture event, and this activation must be achieved *before passivation at the crack tip is completed*. The stipulation is required by the assumption that when crack advance ceases, i.e. passivation is complete, there is no further increase in crack-tip strain. The amount of crack advance (or, faradaically, the average charge density passed) in unit time over the whole crack tip will depend on the electrochemical conditions, e.g. potential, chemistry, etc., and also on the local strain rate, due to the observed[127] dynamic equilibrium between the creation of bared surface and effective removal of previously bared surface by passivation[150] (Fig. 9). Thus the requirement for the passage of a minimum charge density, Q_{min}, for crack advance to continue will be controlled not only by the electrochemical

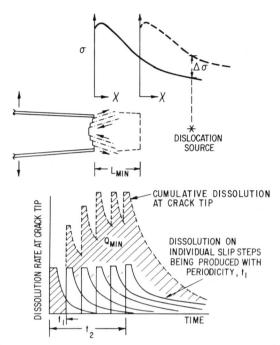

FIG. 9. Illustration of the factors of importance in the constant charge[187] criterion for crack propagation. Propagation will occur only if a dislocation source is activated by the stress increment, $\Delta\sigma$, caused by the crack advance L_{min} associated with the passage of charge density, Q_{min} at the crack tip. This latter amount will be a function of the charge-density increments caused by the individual oxide-rupture events at the crack tip, the periodicity between their occurrence, t_1, and time, t_2, during which the plasticity is occurring.

conditions affecting the dissolution and passivation rates, but also the synergistic relationship between the attainment of both a critical crack-tip strain and minimal strain rate.

The requirement for crack initiation of a critical strain rate *and* a critical strain at the crack tip[196–198] has been experimentally demonstrated in several ductile alloy/aqueous environment systems, and can have important practical consequences. For instance, the prediction of σ_{TH} values based on creep data[188,199] and interrelated metallurgical factors, e.g. strain aging, the understanding as to why propagating cracks can arrest due to stress relaxation[195] if the initial propagation rate is insufficient to maintain a minimal creep rate at the crack tip, and the fact that the $\sigma_{TH}{}^{200}$ or $K_{ISCC}{}^{197,201,202}$ values for ductile alloys will alter with dynamic loading (referred to later in the context of the corrosion fatigue/stress corrosion interaction).

The crack propagation rate, following initiation, is affected by the deformation rate at the crack tip (Fig. 7) through its effect on the periodicity of oxide rupture. According to Vermilyea's original argument[48] the periodicity was dependent on the plastic strain rate at the crack tip (which will also affect the hydrogen-transport kinetics as indicated in Fig. 7). This strain rate may be expected to vary with time, increasing rapidly when the oxide is ruptured and as the underlying strain-hardened material is removed, and decreasing as passivation occurs. Accordingly the oxide rupture periodicity will depend on the time for the increment of strain $\Delta\varepsilon$ through creep to reach the fracture strain of the oxide, ε_f.

Although correct in principle the above model can be modified due to either the definition of plasticity behavior in near-surface locations when the surface may be covered by a different modulus oxide or film, etc., or in the detailed strain criterion for repeated oxide rupture. For instance, the use of uniaxial bulk creep data for the evaluation of the oxide rupture periodicity must be incorrect beyond 'order of magnitude estimations' in view of the highly localized flow adjacent to the crack tip, the range of possible dislocation morphologies, and the known dependency of creep on surface dissolution,[203-206] adsorption,[208,209] surface films,[210,211] periodic stress relaxation,[207] etc. Indeed a direct observation has been made between stress corrosion susceptibility and near-surface mechanical properties,[212] but such relationships have not been widely investigated in aqueous environments.

However, modifications have been made to the strain rate criterion for repeated oxide rupture, which do have experimental justification. For instance Vermilyea[105] modified his original uniaxial creep-controlled criterion,[48] through incorporation of the fact that the activation stress for exhaustion creep will be decreased as the stress at a dislocation source increases when the crack moves forward.

As pointed out above in relation to the crack initiation criterion, Scully[187] attached a similar importance to crack advance and its effect on the activation of dislocation sources. The oxide rupture periodicity in this case was determined by the time required for the crack to advance a constant amount, i.e. the passage of a 'constant-charge density' at the crack tip.

Superficially these two modifications lead to similar predictions of the crack propagation rate/electrode potential relationship, although differences exist in detail.[170] Both Newman,[167] investigating a low alloy steel/caustic system, and Ford,[170] with the 304 stainless steel/Na_2SO_4 system, measured charge density/time relationships and obtained reasonable predictability of the crack propagation rate/electrode potential relationships using a constant-charge criterion with confirmation that the charge passed (or crack advance) for each oxide-rupture event was a constant which could possibly be related to a metallurgical feature (e.g. in 304 stainless steel the calculated crack advance for each oxide rupture event was 0.7 μm, correlating with the dislocation network diameter and the separation of 'arrest' markings on the fracture surface).[213] In addition this constant charge criterion[170,187] offers a qualitative explanation of the propagation rate/stress intensity relationship (Fig. 3); in Stage I the crack velocity increases with stress intensity (and associated strain rate) since the time to penetrate L_{min} (Fig. 9) decreases, whereas in Stage II propagation, L_{min} can be achieved in time periods during which the dissolution rate at the crack tip is constant due to liquid diffusion limitations and the resultant crack velocity becomes stress-intensity (or strain-rate) independent.

Although the above theoretical treatments give a semiquantitative understanding of the effect of oxide rupture and subsequent oxide rupture rate on both the electrode potential and stress dependence of crack propagation, they also indicate that there should be no difference in the *mechanism* of crack advance when the oxide is being ruptured via creep mechanisms under static load or by either applied constantly-rising or cyclic strains, although the kinetics of the rate-determining step may be altered. (For instance Newman[214] has suggested that Q_{min} in the constant-charge criterion will be decreased on the application of a rising load.) In support of this suggestion, several studies [190,191,194] have indicated an increasing crack-propagation rate with rising strain rate although direct quantitative comparison between these dynamic responses and the static V/K relationships (Fig. 3) are not possible because of a lack of accurate knowledge of the crack-tip strain rate in the two situations. Vermilyea and Diegle[215] have demonstrated, however, that quite apart from any chemical limitations to the maximum propagation rate due to activation or diffusional considerations, the crack propagation rate at high strain rates will be limited by the onset of plastic blunting of the crack tip if the ratio (propagation rate:crack-tip strain rate) falls below a critical value, i.e. the flanks of the crack tip are extended more in the tensile axis direction in unit time than the crack tip advances normal to the tensile axis.

Although crack propagation may be limited by plastic blunting under a constantly rising load situation, such a limitation will be less severe under cyclic loading, i.e. corrosion fatigue, since mechanical 'sharpening' can occur. It is possible, however, that in the corrosion fatigue situation that an upper limit in crack growth *per cycle* (da/dN) exists at $CTOD/2$ (where CTOD is the crack tip opening displacement);[216] this limitation is dictated by chemical blunting considerations, for if dissolution at the crack tip gave $da/dN > CTOD/2$, the crack tip would lose its ability to completely resharpen during unloading.

Despite these mechanical and chemical limitations, high environmentally controlled crack propagation rates $(da/dt)_E$ are observed[196,201] at high cyclic loading rates (or short loading times) (Fig. 10(a)) even when the stress intensity amplitude is small (Fig. 10(b)), i.e. a situation symptomatic of small superimposed vibrations on a high mean stress ($R = 0.7$–0.9). It is apparent from Fig. 10(a) and (b) that not only is the environmentally controlled component of the crack growth rate increased during dynamic loading from that observed under static load, but the apparent K_{ISCC} value is decreased. Similar loading rate effects on σ_{TH}[200] or K_{ISCC}[197,201,202] have been observed in other systems, with the minimum 'K_{ISCC}' value under 'zero to maximum load' stressing conditions $(R = 0)$ being obtained where the value K_{max}/Y is (2.7 ± 0.3) $10^{-5} \, \mathrm{m}^{-1/2}$;[217,218] Y in this ratio is the Young's elastic modulus. It has been speculated[196] that this limiting K_{MAX}/Y condition at high loading rates corresponds to the achievement of a strain condition at the crack tip sufficient to rupture the oxide, and that the strain-rate limiting condition in the synergistic $\dot{\varepsilon}/\varepsilon$ initiation criterion is no longer required.

These observations question the validity of predictive models of corrosion fatigue, where the $da/dN/\Delta K$ relationships may be obtained by addition of an 'environmental component' to the $(da/dN)_{\Delta K}$ value obtained in inert environments. In these models the environmental component may be evaluated by integrating the $V/K\dot{\varepsilon}$ relationships over the periodic time of cyclic loading. Such a procedure yields predicted $da/dN/\Delta K$ responses which have been observed and categorized[218] as either 'true corrosion fatigue,' when the environmental effects are observed at $K_{MAX} \leq K_{ISCC}$ (static), or 'stress-corrosion fatigue,' when the environmental effects are confined to $K_{MAX} \geq K_{ISCC}$ (static). Such rigid categorizations are questionable in view of previous discussions of ductile alloy aqueous environment systems since the environmentally controlled propagation rates and (possibly) mechanisms change con-

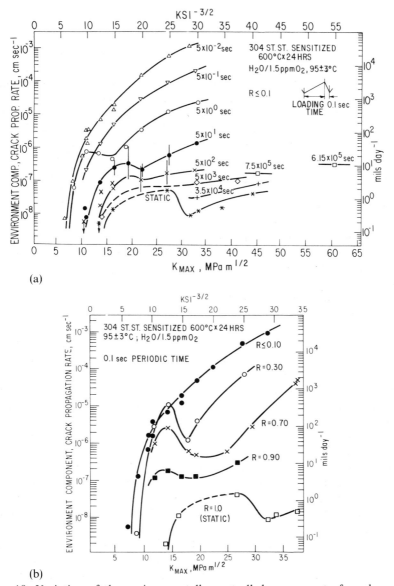

(a)

(b)

FIG. 10. Variation of the environmentally controlled component of crack propagation rate $(da/dt)_E$ with maximum stress intensity for the 304 stainless steel/H_2O system at 95 °C under both static and cyclic loading conditions. [196,201] (a) Effect of loading time under cyclic loading at $R = 0.1$. Data points marked (□) were obtained in a monotonic rising load test, while those marked (*) were obtained under constant load. (b) Effect of R value for static and cyclic loading at 10 Hz frequency. (R = minimum stress intensity/maximum stress intensity.)

tinuously with strain rate, there being no discontinuous change at K_{ISCC} (static). Of more importance, however, it is apparent that predictions of corrosion fatigue behavior based either on superposition models involving the addition[219] of the crack growth rate observed *under static load* to that observed under 'dry' fatigue conditions or on process-competition models[220] based on the choice of the fastest of these two rates, will only be valid for systems in which the rate-determining step for the environmental component of crack velocity, $(da/dt)_E$, is independent of strain rate; for instance high strength alloys in gaseous hydrogen where the rate-determining step in the hydrogen embrittlement model is hydrogen adatom association, surface diffusion[221] or matrix diffusion.[222-224] These 'static' superposition models will not be valid for ductile alloy/aqueous environment systems, because of the strain-rate dependence of $(da/dt)_E$, and will underestimate the actual $(da/dN)_{\Delta K}$ values especially at K_{MAX} values $< K_{ISCC}$ (static).[196]

2.3 Outstanding Mechanistic Questions

The most outstanding question in mechanistic knowledge is the accurate definition of the crack tip conditions, in relation to both electrode potential and liquid composition. Although the topic of localized corrosion has been thoroughly reviewed[225] and although theoretical and experimental investigations[116-119,173-176,226-232] have shown that the crack-tip chemistries and potentials can be markedly different from the values at the exposed specimen surface, the precise predictive analyses and experimental confirmations are few.

Similarly many questions remain on the details of the mechanisms of crack advance: for instance, details of the plasticity processes at the crack tip and how these are affected by surface reactions and plastic constraint, and quantitative details of the alternative hydrogen embrittlement mechanisms. Further aspects are the lack of high temperature thermodynamic and kinetic data for extrapolation of the reasonably well-defined mechanistic theory at lower temperatures to high temperatures; and details of the microscopic compositional variations at, e.g. grain boundaries, and how these affect the crack-tip electrode reactions. Finally, there is the need for better understanding of the fractographic aspects of environmentally controlled fracture, for although crack path changes may be understood in terms of the path down which crack propagation by a given mechanism is most energetically favorable or is fastest, the specific details of the arguments remain quantitatively hazy.

3. DEVELOPMENT OF TESTING TECHNIQUES RELEVANT TO PRACTICE

The state of the art for testing techniques, their reproducibility and relevance to practice has been recently reviewed.[233-235] In particular, the interpretation of the uncertainties and irreproducibility of the traditional time to failure tests for smooth specimens under constant load or displacement conditions has been underlined.[233] In addition the un-qualified use of "accelerated' testing techniques (necessitated by long service-life requirements where crack propagation rates of 10^{-8} mm s^{-1} are significant) involving high temperatures, high stress, oxidizing conditions, etc., may be questionable since, in some cases, these techniques may lead to a change in cracking mechanism, rather than merely an increase in the rate-determining step of the mechanism expected in service.

Recently slow-strain rate tests[235] have become widely accepted as quality control or screening techniques, quite apart from their usefulness in mechanistic studies. The prime justification for this technique is that it accelerates a known rate-determining step in the cracking mechanism of ductile alloy/aqueous environment systems, e.g. oxide-rupture rate; it is not surprising therefore that good correlations are observed between stress corrosion susceptibility rated by this rapid technique and by more protracted methods involving static loading. Limits to its practical usefulness in detecting cracking susceptibility are apparent for systems of marginal susceptibility caused by, for instance, a high passivation rate; in these circumstances very low strain rates (or very long testing times) may be necessary in order to detect any change in ductility parameters due to cracking, and these low strain rates *per se* will tend to give low susceptibility, whereas the detection of cracking in systems exhibiting less tendency for rapid passivation can be achieved at high applied strain rates.[236]

Testing techniques using fracture-mechanics concepts have found in-creasing use, since it can be argued that components in practice normally have manufacturing defects (or, at least, defects cannot be guaranteed to be absent with sizes less than the NDT resolution limit). In addition, the improvement of crack-following techniques has enabled crack growth rates of the order of 10^{-8} mm s^{-1} to be measured, thereby both reducing the necessity for accelerative methods and providing test data which can be incorporated directly into a design criterion.

The application of the fracture mechanics approach to design concepts[102,237-242] relies on the definition of the boundary lines on the stress/crack-depth diagram between no crack growth, subcritical crack growth and catastrophic failure (or plastic collapse) (Fig. 11). These boundaries are experimentally definable with limits being given by σ_{TH} and σ_{UTS} for initially smooth specimens and K_{ISCC} and K_{IC} for cracked samples. Whether or not a component will crack from the time of commissioning will depend on the tensile stress/pre-existing defect size combination; if the initial stress intensity associated with this combination is greater than K_{ISCC}, than the component lifetime or inspection periodicity

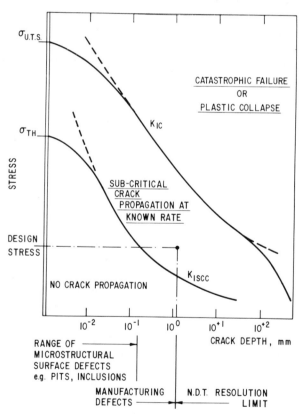

FIG. 11. Tensile stress/crack depth diagram, illustrating the regions of 'no crack growth', 'subcritical crack growth' and 'catastrophic failure' (or plastic collapse), and how this diagram may be used for the assessment of the likelihood of cracking in a defective component.

may be determined by integration of the appropriate velocity/stress-intensity curve.

Although such a design approach is powerful in principle and operation, questions of detail remain as to its unqualified practical use. For instance, the precise definition of the σ_{TH}/K_{ISCC} boundary at small crack lengths has not been quantified theoretically (as has been done for the σ_{UTS}/K_{IC} boundary[243] nor is this boundary amenable to precise experimental determination, especially when K_{ISCC} is pragmatically defined at very low propagation rates. Further the locus of the σ_{TH}/K_{ISCC} boundary should (theoretically) change with the degree of plastic constraint, a condition commonly encountered in practice, and, as discussed above, the position of this boundary for ductile alloy/aqueous environment systems will change with dynamic loading. This latter point is of interest, not only from the question of the validity of rising load K_{ISCC} testing[244] for *ductile* alloys but also from the practical design aspect,[102] for although a component may be designed so that it will not crack under static load conditions, it may well do so during transient loading situations.

4. SUMMARY

It is apparent in reviewing the current state of mechanistic understanding of cracking in ductile alloy/aqueous environment systems that, although many questions remain unanswered (especially in the definition of the precise metallurgical and environmental conditions at the crack tip), the basic thermodynamic and kinetic requirements for crack initiation and propagation are understood and are quantifiable for the main alternative mechanisms. Consequently the dangers of cracking in practical situations can be assessed from a mechanistic base as well as from the existing foundation of 'empirical' data.[245]

It is also apparent that although the current arguments concerning the validity of the strain-assisted active-path or hydrogen-embrittlement mechanisms are important academically, they are probably of secondary importance from a practical viewpoint since the *rate-controlling* reactions, oxide rupture, passivation and liquid diffusion, can be common to both competing theories (Fig. 7) especially for ductile alloy/aqueous environment systems. Thus cracking susceptibility in these systems may be understood from knowledge of the composite behaviors of the rate-determining steps and how they are affected by such system variables as electrode potential, dislocation morphology, anion content, etc., which are amenable to design changes.

Finally it is apparent that classification of 'stress corrosion', 'hydrogen embrittlement,' 'true corrosion fatigue' and 'stress corrosion fatigue' can be artificial and confining in terms of remedial actions. As well as the spectrum of conditions between 'active-path', 'stress-assisted active-path' and hydrogen-embrittlement mechanisms proposed by Parkins[16] there is a similar spectrum of behavior between static (stress corrosion) and dynamic (corrosion fatigue) loading conditions.

ACKNOWLEDGEMENTS

Acknowledgment is given for encouragement and discussions with former colleagues at the Central Electricity Research Laboratories, Leatherhead, UK and with present colleagues at General Electric, USA. Particular acknowledgement is extended to R. Carter who gave comments on the manuscript and to the Electric Power Research Institute who gave permission for publication of some of the data.

REFERENCES

1. S. P. Lynch and N. E. Ryan, *Proc. 2nd Int. Congress on Hydrogen in Metals*, Paris, 1977.
2. A. S. Tetelman and S. Kunz, Conference on Stress Corrosion Cracking and Hydrogen Embrittlement of Iron-Base Alloys, Firminy, June 1973, p. 359, eds R. W. Staehle, J. Hochmann, R. D. McCright, J. E. Slater, NACE, Houston (1977).
3. W. Frank and L. Graf, Conference on Surface Effects in Crystal Plasticity, Hohegeiss, Germany, September 1975, p. 781, Eds R. M. Latanision and J. T. Fourie, Noordhof-Leyden (1977).
4. R. W. Staehle, Conference on Mechanisms of Environment Sensitive Cracking of Materials, University of Surrey, April 1977, Eds P. R. Swann, F. P. Ford and A. R. C. Westwood, Metals Society, p. 574.
5. *Environmental-sensitive mechanical behavior*, Baltimore, Maryland, June 1965, Eds ARC Westwood and N. S. Stoloff, Gordon and Breach.
6. *Fundamental aspects of stress-corrosion cracking*, Ohio State University, Sept. 1967, Eds R. W. Staehle, A. J. Forty, D. Van Rooyen. NACE, Houston.
7. *Theory of stress-corrosion cracking*, Ericiera, Portugal, March 1971, Ed. J. C. Scully, NATO, Brussels.
8. *Stress-corrosion cracking and hydrogen embrittlement of iron-base alloys*,

Firminy, France, June 1973, Eds R. W. Staehle, J. Hochmann, R. D. McCright and J. E. Slater, NACE, Houston (1977).

9. *L'Hydrogene dans les Metaux*, Paris 1972, Ed. M. P. Bastien, Science et Industrie.

10. *Hydrogen in metals*, Eds I. M. Bernstein and A. W. Thompson, ASM (1973).

11. *Effect of hydrogen on behavior of materials*, Jackson Lake, Wyoming, September 1975, Eds A. W. Thompson and I. M. Bernstein, AIME.

12. *Surface effects in crystal plasticity*, Hohegeiss, Germany, September 1975, Eds R. M. Latanision and J. T. Fourie, Noordhof-Leyden (1977).

13. *Mechanisms of environment sensitive cracking of materials*, University of Surrey, UK, April 1977, Eds P. R. Swann, F. P. Ford and A. R. C. Westwood, The Metals Society.

14. *Corrosion fatigue-chemistry, mechanics and microstructure*, University of Connecticut, Storrs, June 1971, Eds O. Devereaux, A. J. McEvily, R. W. Staehle, NACE, Houston (1972).

15. *Corrosion-fatigue*, University of Newcastle, April 1979, Metal Science Vol. 13, 1979.

16. R. N. Parkins, *Br. Corr. J.*, **7**, 15 (1972).

17. E. H. Dix, *Trans AIME*, **11**, 137 (1940).

18. R. B. Mears, R. H. Brown and E. H. Dix, *Symposium on Stress Corrosion Cracking of Metals*, ASTM/AIME, Philadelphia, p. 329 (1944).

19. E. H. Dix, *Trans ASM*, **42**, 1057 (1950).

20. J. M. Krafft and J. H. Mulherin, *Trans ASM*, **62**, 64 (1969).

21. T. P. Hoar and J. G. Hines, *Proc 8th Meeting CITCE*, Madrid, 1956, Butterworths (1958).

22. J. G. Hines and T. P. Hoar, *J. Appl. Chem.*, **8**, 764 (1958).

23. J. G. Hines and T. P. Hoar, *J. Iron and Steel Inst.*, **182**, 124 (1956).

24. T. P. Hoar and J. M. West, *Nature*, **181**, 35 (1958).

25. J. G. Hines, *Corrosion Sci.*, **1**, 21 (1961).

26. T. P. Hoar and J. M. West, *Proc. Roy. Soc.*, **A268**, 304 (1962).

27. T. P. Hoar, *Corrosion*, **19**, 331 (1963).

28. T. P. Hoar, Ref. 6, p. 98.

29. P. R. Swann, *Corrosion*, **19**, 102t (1963).

30. P. R. Swann and J. D. Embury, *High strength materials*, ed. V. F. Zachay, p. 327, J. Wiley, New York (1965).

31. J. M. Silcock and P. R. Swann, Ref. 13, p. 66.

32. G. Thomas and J. Nutting, *J. Inst. Metals*, **88**, 81 (1959/60).

33. W. D. Robertson and A. S. Tetelman, *Strengthening mechanisms in solids*, p. 217, ASM, Cleveland (1962).

34. D. Tromans and J. Nutting, *Corrosion*, **21**, 143 (1965).

35. E. N. Pugh, Ref. 5, p. 351.

36. P. R. Swann and J. Nutting, *J. Inst. Met.*, **88**, 478 (1960).

37. D. L. Douglass, G. Thomas and W. R. Roser, *Corrosion*, **20**, 15t (1964).

38. H. W. Pickering and P. R. Swann, *Corrosion*, **19**, 373t (1963).

39. T. Smith and R. W. Staehle, *Corrosion*, **23**, 117 (1967).

40. F. A. Champion, Symposium on Internal Stresses in Metals and Alloys, p. 468, London 1948, Inst. of Metals.

41. H. L. Logan, *J. Research N.B.S.*, **48**, 99 (1952).

42. R. W. Staehle, Ref. 7, p. 223.
43. T. R. Beck, *Corrosion*, **30**, 408 (1974).
44. J. C. Scully, *Corrosion Sci.*, **7**, 197 (1967).
45. J. C. Scully, *Corrosion Sci.*, **8**, 513 (1968).
46. J. C. Scully, *Corrosion Sci.*, **8**, 771 (1968).
47. T. P. Hoar, Ref. 7, p. 106.
48. D. A. Vermilyea, *J. Electrochem. Soc.*, **119**, 405 (1972).
49. A. J. Forty and P. Humble, *Phil. Mag.*, **8**, 247 (1963).
50. A. J. McEvily and A. P. Bond, *J. Electrochem. Soc.*, **112**, 131 (1965).
51. A. J. Sedriks, P. W. Slattery, E. N. Pugh, *Trans ASM*, **62**, 238 (1969).
52. E. N. Pugh, Ref. 7, p. 418.
53. A. J. McEvily and A. P. Bond, *Trans ASM*, **60**, 661 (1967).
54. T. R. Beck, Ref. 7, p. 64.
55. H. H. Pickering, Ref. 6, p. 159.
56. R. N. Parkins, *Br. Corr. J.*, **14**, 5 (1979).
57. *Hydrogen damage*, ed. C. D. Beachem, ASM (1977).
58. J. P. Hirth and H. H. Johnson, *Corrosion*, **32**, 3 (1976).
59. N. J. Petch and P. Stables, *Nature*, **169**, 842 (1952).
60. N. J. Petch, *J. Iron Steel Inst.*, **174**, 25 (1955).
61. N. J. Petch, *Phil. Mag.*, **1**, 331 (1956).
62. E. G. Coleman, D. Weinstein, W. Rostoker, *Acta Met.*, **9**, 491 (1961).
63. R. B. Heady, *Corrosion*, **33**, 441 (1977).
64. A. A. Griffith, *Phil. Trans. Roy. Soc.* (London), **A221**, 163 (1920).
65. H. Nichols and W. Rostoker, *Acta Met.*, **9**, 504 (1961).
66. M. Kamdar and A. R. C. Westwood, Ref. 5, p. 581.
67. G. F. Old and P. Trevena, *3rd. Int. Conf. on Mechanical Behavior of Materials*, Cambridge, August 1979, p. 397, eds K. J. Miller and R. F. Smith, Pergamon Press.
68. H. H. Uhlig, *Physical metallurgy of stress corrosion fracture*, ed. T. N. Rhodin, p. 11, Interscience, NY (1959).
69. A. R. Troiano, *Trans ASM*, **52**, 54 (1960).
70. R. A. Oriani, *Berichte der Bunsenges für Phys. Chem.*, **76**, 848 (1972).
71. R. A. Oriani and P. H. Josephic, *Acta Met.*, **22**, 1065 (1974).
72. R. A. Oriani, Ref. 8, p. 351.
73. K. Yoshino and C. J. McMahon, *Met. Trans.*, **5**, 363 (1974).
74. R. Raj and V. K. Varadan, Ref. 13, p. 426.
75. P. Doig and G. T. Jones, Ref. 13, p. 446.
76. W. W. Gerberich, Y. T. Chen, C. St. John, *Met. Trans.*, **6**, 1485 (1975).
77. J. F. Lessar and W. W. Gerberich, *Met. Trans.*, **7A**, 953 (1976).
78. F. de Kazinsky, *J. Iron Steel Inst.*, **177**, 85 (1954).
79. C. A. Zapffe and M. Haslem, *Metals Tech.*, **13**, 1 (1946).
80. C. A. Zapffe and C. E. Sims, *Trans AIME*, **145**, 225 (1941).
81. A. J. Tetelman and W. D. Robertson, *Trans AIME*, **224**, 775 (1962).
82. A. J. Tetelman, *Fracture of solids*, eds D. C. Drucker and J. Gilman, p. 671, J. Wiley, New York (1963).
83. R. A. Gilman, Ref. 8, p. 326.
84. D. N. Williams, *J. Inst. Metals*, **91**, 147 (1962).
85. J. C. Scully and D. T. Powell, *Corrosion Sci.*, **10**, 719 (1970).

86. R. B. Benson, D. K. Dunn and L. W. Roberts, *Trans AIME*, **242**, 2199 (1968).
87. R. A. McCoy and W. W. Gerberich, *Met. Trans.*, **4**, 539 (1973).
88. R. A. McCoy, Ref. 10, p. 169.
89. R. J. Asaro, A. J. West, W. A. Tiller, Ref. 8, p. 1115.
90. S. S. Birley and D. Tromans, *Corrosion*, **27**, 63 (1971).
91. C. D. Beachem, *Met. Trans.*, **3**, 437 (1972).
92. C. D. Beachem, Ref. 8, p. 376.
93. M. R. Louthan, G. R. Gaskey, J. A. Donovan and D. E. Rawl, *Mats. Sci. Eng.*, **10**, 357 (1972).
94. M. R. Louthan, J. A. Donovan and D. E. Rawl, *Corrosion*, **29**, 108 (1973).
95. A. W. Thompson and I. M. Bernstein, The role of metallurgical variables in hydrogen assisted environmental fracture. Rockwell Science Center Report SC-PP-75-63. (To be published in *Advanced Corrosion Science and Technology.*)
96. J. K. Tien, A. W. Thompson, I. M. Bernstein, and R. J. Richards, *Met. Trans.*, **7A**, 821 (1976).
97. J. K. Tien, R. J. Richards, O. Buck and H. L. Marcus, *Scripta Met.*, **9**, 1097 (1975).
98. A. W. Thompson, Ref. 10, p. 91.
99. S. M. Weiderhorn, *Materials Science Research*, **3**, 503 (1966).
100. T. P. Hoar and G. P. Rothwell, *Electrochim. Acta*, **15**, 1037 (1976).
101. J. M. West, *Metal. Sci. J.*, **7**, 169 (1973).
102. F. P. Ford, as Ref. 67, p. 431.
103. Orowan, *Trans. Inst. Engrs. Shipbuilders*, Scotland, **89**, 165 (1945).
104. G. R. Irwin, 9th Int. Cong. Appl. Mech. VIII paper 101, University of Brussels, 1957.
105. D. A. Vermilyea, Ref. 8, p. 208.
106. A. J. Kransz, *Int. J. Fracture.*, **14**, 5 (1978).
107. E. R. Fuller and R. M. Thomson, as Ref. 67, p. 485.
108. I. C. Howard, as Ref. 67, p. 463.
109. W. W. Gerberich and Y. T. Chen, *Met. Trans.*, **6A**, 271 (1975).
110. R. A. Oriani, Ref. 6, p. 32.
111. W. W. Gerberich and C. D. Hartblower, Ref. 6, p. 420.
112. C. St. John and W. W. Gerberich, *Met. Trans.*, **4**, 589 (1973).
113. R. N. Parkins, N. J. H. Holroyd and R. R. Fessler, *Corrosion*, **34**, 253 (1978).
114. B. W. Roberts and P. Greenfield, *Corrosion*, **35**, 402 (1979).
115. M. O. Speidel, *Corrosion*, **32**, 187 (1976).
116. R. Picinini, M. Marek, A. J. Pourbaix, and R. F. Hochman, *Localized corrosion*, Williamsburg, December 1971, p. 179, Eds R. W. Staehle, B. F. Brown and J. Kruger, NACE, Houston (1974).
117. J. A. Davis, as Ref. 116, p. 168.
118. J. A. Smith, M. H. Peterson and B. F. Brown, *Corrosion*, **26**, 539 (1970).
119. B. F. Brown, C. T. Fujii and E. P. Dahlberg, *J. Electrochem. Soc.*, **116**, 218 (1969).
120. C. Tyzack, *Br. Corr. J.*, **7**, 268 (1972).
121. C. Tyzack, Ref. 14, p. 319.

306   F. P. FORD

122. C. Tyzack, *Br. Corr. J.*, **6**, 219 (1971).
123. C. Tyzack, *Br. Corr. J.*, **7**, 189 (1972).
124. C. Tyzack, *Int. J. Press. Ves. and Piping*, **2**, 231 (1974).
125. T. P. Hoar and R. W. Jones, *Corrosion Sci.*, **13**, 725 (1973).
126. T. P. Hoar and J. R. Galvale, *Corrosion Sci.*, **10**, 211 (1970).
127. T. P. Hoar and F. P. Ford, *J. Electrochem. Soc.*, **120**, 1013 (1973).
128. T. P. Hoar and J. C. Scully, *J. Electrochem. Soc.*, **111**, 348 (1964).
129. T. R. Beck, *J. Electrochem. Soc.*, **115**, 880 (1968).
130. D. J. Lees and T. P. Hoar, European Federation of Corrosion, *Electrochemical methods for stress-corrosion testing*, Firminy, September 1978.
131. F. P. Ford, Ref. 13, p. 125, *Metal Science*, 326 (July 1978).
132. T. Hagyard and W. B. Earl, *J. Electrochem. Soc.*, **114**, 694 (1967).
133. N. D. Tomashov and L. P. Verastinina, *Electrochim. Acta*, **15**, 501 (1976).
134. J. R. Ambrose and J. Kruger, *Corrosion*, **28**, 30 (1972).
135. J. O'M Bockris and A. K. N. Reddy, *Modern electrochemistry*, Vol. 2, p. 1338, Plenum (1970).
136. M. O. Speidel, Ref. 7, p. 289.
137. J. Berggreen, PhD Thesis, University of Erlangen-Nurnberg 1973.
138. I. M. Bernstein and A. W. Thompson, *Alloy and microstructural design*, eds J. K. Tien and G. S. Ansell, Academic Press, New York (1976).
139. P. R. Rhodes, *Corrosion*, **25**, 462 (1969).
140. A. W. Thompson and I. M. Bernstein, *Reviews on Coatings and Corrosion*, **2**, 3 (1975).
141. H. P. Van Leeuwen, *Corrosion*, **29**, 197 (1973).
142. H. P. Van Leeuwen, J. A. M. Boogers and C. J. Stentler, *Corrosion*, **31**, 23 (1975).
143. R. J. H. Wantill, *Corrosion*, **31**, 143 (1975).
144. D. J. Lees, F. P. Ford and T. P. Hoar, *Metals and Materials*, **7**, 5 (1973).
145. J. C. Scully, Ref. 13, p. 1.
146. T. Shibata and R. W. Staehle, Strain electrochemistry of passive iron, *Corrosion-71*, March 1971, Chicago.
147. J. R. Ambrose and J. Kruger, *J. Electrochem. Soc.*, **121**, 599 (1974).
148. H. Leidheiser and E. Kellerman, *Corrosion*, **26**, 99 (1970).
149. B. E. Wilde, *J. Electrochem. Soc.*, **118**, 1717 (1971).
150. T. Murata and R. W. Staehle, Report No. C00 1319–71 (Q19), Ohio State University Research Foundation, Columbus, Ohio, 1971.
151. G. Theus and R. W. Staehle, Ref. 8, p. 845.
152. M. R. Louthan and R. G. Derrick, *Corrosion Sci.*, **15**, 565 (1975).
153. C. L. Huffman and J. M. Williams, *Corrosion*, **16**, 430 (1960).
154. J. C. Sherlock and L. L. Shreir, *Corrosion Sci.*, **10**, 561 (1970).
155. J. C. Sherlock and L. L. Shreir, *Corrosion Sci.*, **11**, 543 (1971).
156. R. W. Staehle, J. J. Royuela, T. L. Raredon, E. Serrate, C. R. Morin and R. V. Farrar, *Corrosion*, **26**, 451 (1970).
157. H. H. Lee and H. H. Uhlig, *J. Electrochem. Soc.*, **117**, 18 (1970).
158. R. T. Newberg and H. H. Uhlig, *J. Electrochem. Soc.*, **119**, 981 (1972).
159. H. H. Uhlig and E. W. Cook, Jr., *J. Electrochem. Soc.*, **116**, 173 (1969).
160. P. T. Gilbert and S. E. Hadden, *J. Inst. Metals*, **77**, 237 (1950).

161. W. Williams and J. Eckel, *J. Am. Soc. Nav. Eng.*, **68**, 93 (1956).
162. H. H. Uhlig, *Corrosion and corrosion control*, 2nd edn, p. 314, Wiley (1971).
163. M. J. Humphries and R. N. Parkins, *Corrosion Sci.*, **7**, 747 (1967).
164. D. Van Rooyen, 1st Int. Conf. on Stress Corrosion Cracking, London 1962, Butterworths (1962).
165. G. Lafranconi, F. Mazza, E. Sivieri and S. Torchio, *Corrosion Sci.*, **18**, 617 (1978).
166. R. Greenwood and J. C. Scully (in preparation), J. C. Scully, Ref. 13, p. 1.
167. J. F. Newman, CEGB Report RD/L/N78/75, August 1975; Ref. 13, p. 19.
168. R. E. Stoltz and R. N. Pelloux, *Corrosion*, **29**, 13 (1973).
169. F. Zucchi, G. Trabanelli, A. Frignani and M. Zucchini, *Corrosion Sci.*, **18**, 87 (1978).
170. F. P. Ford and M. Silverman, *Corrosion*, **36**, 558 (1980).
171. J. R. Galvele, S. B. deWexler and I. Gardiazabal, *Corrosion*, **31**, 352 (1975).
172. J. A. Feeney and M. J. Blackburn, Ref. 7, p. 355.
173. T. R. Beck and E. H. Grens, *J. Electrochem. Soc.*, **116**, 117 (1969).
174. R. R. Shuck and J. L. Swedlow, as Ref. 116, p. 190.
175. R. R. Shuck and J. L. Swedlow, as Ref. 116, p. 208.
176. R. J. Taunt and W. Charnock, *Mat. Sci. Eng.*, **35**, 219 (1978).
177. G. M. Scamens and P. R. Swann, Ref. 8, p. 166.
178. R. M. Latanision and R. W. Staehle, Ref. 6, p. 214.
179. A. W. Thompson, *Mat. Sci. Eng.*, **14**, 253 (1974).
180. R. R. Vandervoort, A. W. Ruotola, and E. L. Raymond, *Met. Trans.*, **4**, 1175 (1973).
181. R. T. Ault, K. O. McDowell, P. L. Hendricks and T. M. F. Ronald, *Trans. ASM*, **60**, 79 (1967).
182. M. O. Speidel, Ref. 6, p. 561.
183. A. Kelly and R. B. Nicholson, *Prog. Mats. Sci.*, **10**, 149 (1963).
184. N. E. Paton and A. W. Sommer, *Proc. 3rd Int. Conf. Strengths of Metals and Alloys*, Cambridge, Vol. 1, p. 101, Inst. of Metals (1973).
185. K. C. Thomas, R. Stickler and R. Allio, *Corrosion Sci.*, **5**, 71 (1965).
186. M. N. Saxena and R. A. Dodd, Ref. 5, p. 455.
187. J. C. Scully, *Corrosion Sci.*, **15**, 207 (1975).
188. R. N. Parkins, ASTM STP 665, p. 5, eds G. M. Ugiansky and J. H. Payer (1977).
189. R. N. Parkins, *J. Strain Analysis*, **10**, 251 (1975).
190. T. Adepoju and J. C. Scully, *Corrosion Sci.*, **15**, 415 (1975).
191. J. C. Scully and T. Adepoju, *Corrosion Sci.*, **17**, 789 (1977).
192. J. C. Scully, as Ref. 188, p. 237.
193. J. M. Sutcliffe, R. R. Fessler, W. K. Boyd and R. N. Parkins, *Corrosion*, **28**, 313 (1972).
194. M. Kermani and J. C. Scully, *Corrosion Sci.*, **89**, 19 (1979).
195. W. R. Wearmonth, G. P. Dean, and R. N. Parkins, *Corrosion*, **29**, 251 (1973).
196. F. P. Ford and M. Silverman, Second Semiannual Report on EPRI Contract RP1332-1. Mechanisms of environmentally-enhanced cracking, December 1979.

197. R. N. Parkins and B. S. Greenwell, *Metal Sci.*, 465 (August 1977).
198. R. N. Parkins, G. P. Marsh and J. T. Evans, Strain rate effects in environment sensitive fracture. Paper at US–Japan meeting on Predictive Corrosion Testing in LWRs, Mount Fuji, Japan, June 1978.
199. R. N. Parkins, 6th Symposium on Line Pipe Research, Am. Gas Assoc., Cat. No. L30175, Q1-22 (1979).
200. R. R. Fessler, *Pipe Line Ind.*, **44**, 37 (1976).
201. F. P. Ford and M. Silverman, *Corrosion*, **36**, 597 (1980).
202. D. Dawson and R. N. Pelloux, *Met. Trans.*, **5**, 723 (1974).
203. H. H. Uhlig, *J. Electrochem. Soc.*, **123**, 1699 (1976).
204. R. W. Revie and H. H. Uhlig, *Acta Met.*, **22**, 619 (1974); *Scripta Met.*, **8**, 1231 (1974).
205. E. W. Hart, *Proc. 14th Sagamore Army Materials Res. Conf.*, p. 210, Syracuse Press (1966).
206. D. J. Duquette, H. Hahn and P. Andresen, Ref. 12, p. 469.
207. J. T. Evans and R. N. Parkins, *Acta Met.*, **24**, 511 (1976).
208. A. Pfutzenreuter and G. Mazing, *Z. Metallk.*, **42**, 361 (1951).
209. R. M. Latanision, H. Opperhauser and A. R. C. Westwood, Science of hardness testing and its research applications. ASM, p. 432 (1973).
210. G. E. Ruddle and H. G. F. Wilsdorf, *Appl. Phys. Letters*, **12**, 271 (1968).
211. V. K. Sethi and R. Gibala, *Scripta Met.*, **9**, 527 (1975).
212. I. R. Kramer, *Corrosion*, **31**, 383 (1975); **31**, 391 (1975).
213. N. A. Neilson, Ref. 8, p. 1108.
214. J. F. Newman, CEGB Report RD/L/N120/78, November 1978.
215. D. A. Vermilyea and R. B. Diegle, *Corrosion*, **32**, 26 (1976).
216. B. Tomkins, *Metals Sci.*, **13**, 387 (1979).
217. J. D. Harrison, *Metal Construction and British Welding J.*, **93** (1970).
218. M. O. Speidel, M. J. Blackburn, T. R. Beck and J. E. Feeney, Ref. 14, p. 324.
219. R. P. Wei and J. D. Landes, *Mats. Res. and Standards*, 25 July (1969).
220. I. M. Austen and E. F. Walker, Ref. 13, p. 334.
221. D. P. Williams and H. G. Nelson, *Met. Trans.*, **1**, 63 (1970).
222. D. P. Williams and H. G. Nelson, *Met. Trans.*, **3**, 2107 (1972).
223. E. A. Steigerwald, F. W. Schaller and A. R. Troiano, *Trans AIME*, **215**, 1048 (1959).
224. H. H. Johnson, Ref. 6, p. 439.
225. *Localized corrosion*, Williamsburg December 1971, eds R. W. Staehle, B. F. Brown and J. Kruger, NACE, Houston (1974).
226. B. G. Ateya and H. W. Pickering, *J. Electrochem. Soc.*, **122**, 1018 (1975).
227. E. McCafferty, *J. Electrochem. Soc.*, **121**, 1007 (1974).
228. D. A. Vermilyea and C. S. Tedmon, *J. Electrochem. Soc.*, **117**, 437 (1970).
229. H. W. Pickering and R. P. Frankenthal, *J. Electrochem. Soc.*, **119**, 1297 (1972).
230. P. H. Melville, *Br. Corr. J.*, **14**, 15 (1979).
231. P. Doig and P. E. J. Flewitt, *Proc. Roy. Soc.*, London, **A357**, 432 (1977).
232. D. F. Taylor and M. Silverman, *Corrosion*, **36**, 544 (1980).
233. R. N. Parkins, F. Mazza, J. J. Royuela and J. C. Scully, *Br. Corr. J.*, **7**, 154 (1972).
234. Stress-corrosion—new approaches, ASTM STP 610, ed. H. L. Craig (1976).

235. Stress-corrosion cracking—the slow strain rate technique, ASTM STP 665, eds G. M. Ugiansky and J. H. Payer (1979).
236. J. C. Scully and D. T. Powell, *Corrosion Sci.*, **10**, 719 (1970).
237. R. B. Diegle, Application of stress-corrosion and corrosion-fatigue data to design. Chapter in *ARPA handbook on stress-corrosion and corrosion fatigue*.
238. M. O. Speidel and M. V. Hyatt, *Stress-corrosion cracking of high-strength alloys*, *Advances in Corrosion Science and Technology*, Plenum Press Vol. 2, 1972.
239. D. O. Sprouls, M. B. Schumaker and I. D. Walsh, Evaluation of stress-corrosion crack susceptibility using fracture-mechanics techniques. Final Report of Contract NAS 8, 21 487, Alcoa (1973).
240. J. G. Kaufman *et al.*, Technical Report AMFL T.R. 69-255, November 1969.
241. P. M. Scott, *Tolerance of flaws in pressurized components, I. Mech. E. Conference* May 1978 London, p. 77, Mechanical Engineering Publications, London (1978).
242. T. A. Prater, F. P. Ford and L. F. Coffin, *Metal Sci.*, 424 (August 1980).
243. P. T. Heald, G. M. Spink and P. J. Worthington, *Mats. Sci. Eng.*, **10**, 129, (1972).
244. W. G. Clark, Jr. and J. D. Landes, Ref. 234, p. 108.
245. *Handbook of Stress-Corrosion and Corrosion-Fatigue*, ARPA, eds R. W. Staehle and M. O. Speidel (to be published).
246. R. N. Parkins, Ref. 235, p. 5.
247. J. C. Scully, ibid, p. 237.

INDEX